Performance-Based Contracts for Road Projects

Ashish Gajurel

Performance-Based Contracts for Road Projects

Comparative Analysis of Different Types

Springer

Ashish Gajurel
Institution of Engineering
Civil Engineering/Transportaion
Tribhuvan University
Kathmandu, Nepal

ISBN 978-81-322-1301-7 ISBN 978-81-322-1302-4 (eBook)
DOI 10.1007/978-81-322-1302-4
Springer New Delhi Heidelberg New York Dordrecht London

Library of Congress Control Number: 2013944068

Printed on acid-free paper

Springer is part of Springer Science+Business Media (www.springer.com)

Preface

This book is a transformation of my Master's Thesis "Comparison of different forms of performance-based contracts for infrastructure projects", which I authored for the fulfillment of my Masters of Science (M.Sc.) degree in Transportation Systems from the department of Building Process Management and Real Estate Management, Technische Universität München, Germany under the supervision of Univ.-Prof. Dr.-Ing. Josef Zimmermann.

The book is targeted at both academic and professional audiences. It is relevant not only to civil engineering and transportation engineering students, but also to professionals working in the infrastructure and construction industries. The book will also prove useful to public sector employees responsible for development, construction, operation, maintenance, and management of public infrastructure.

The objective of this book is to compare the available performance-based road project delivery systems, which are currently in implementation worldwide. The book recognizes Performance-based Road Management and Maintenance Contract (PMMR), Funktionsbauvertrag (FBV), and Public-Private Partnership (PPP) as the most widely used performance-based road project delivery systems. Infrastructure development is capital-intensive and time-consuming. Therefore, the public sector and the private sector must work together to deliver the required infrastructure in an efficient manner. The purpose of this book is to compare the above-stated delivery systems and identify the similarities and differences between them. Such a comparison will help in selecting suitable methods for projects at hand, based on the prevalent circumstance and conditions. This book attempts to comprehensively cover the three types of delivery systems mentioned above including, concept, procedure, nature, and comparison with the other systems.

PMMR, FBV, and PPP are all output-based contracts, where the required level of performance must be delivered. These contracts do not define the process, procedures, techniques, and materials requirements. This provides the opportunity for the contractor to be innovative in the selection of methods, techniques, and materials. They are long-term contracts, usually 15–30 years. PMMR and PPP are more widely used in Anglo-Saxon countries whereas FBV was developed and is implemented in Germany. PMMR and FBV are developed for road project delivery whereas PPP comprises different infrastructures such as roads, hospitals, water supply, and railway. However, in principle, FBV and PMMR could be extended to other types of projects by developing the clear functional requirements for each type of project.

Since governments have limited resources and a wide range of areas requiring investment, optimal use of funds is extremely important. Hence, public sector organizations are increasingly exploring more efficient and effective construction contracts which ensure low investment costs and higher quality. The book presents three construction contracts which could guide the public sector in using suitable contracts based on their specific requirements. These details are also useful for construction companies to understand contract selection and details of PMMR, FBV, and PPP.

I have attempted to make everything simple and easy to understand with suitable tables, figures, graphs, examples, and case studies. Practical examples of FBV (Road Project: Rastplatz "Blauer Stein"– Miel on the Highway BAB A 61, Germany) and PMMR (PMMR in Argentina) have been presented as case studies in order to understand the utilization of these contracts in the real world.

I would be happy to receive the readers' comments and suggestions for further improvement and enrichment of this book.

Kathmandu, Nepal Ashish Gajurel

Acknowledgment

I would like to express my gratitude to Professor Zimmermann and Mr. Benno Vocke from Technische Universität München, Germany, for all the support, guidance, and inputs provided by them for the book and to me, without which the completion of the book would not have been possible.

I would also like to thank my professors and friends at Technische Universität München, Germany, for all that I have learned about transportation and traffic engineering. I am grateful for the love and encouragement my parents, Murari and Shova Gajurel, and sister Asma provided in every step of my life. I thank my uncles Chet Nath Dahal and Deepak Gajurel and the whole Gajurel family for their continuous support and motivation. Finally, I would like to thank my wife, Ajita Adhikari, for her encouragement, support, and love.

Contents

Abbreviations

A	
A 61	Autobahn 61 (Highway 61)
ASCE	American Society of Civil Engineers
B	
BAB	Bundesautobahn (Federal Highway)
BMVBS	Bundesministerium für Verkehr, Bau und Stadtentwicklung (Federal Ministry of Transport, Building and Urban Development)
BOO/BOOT	Build-own-operate/build-own-operate-transfer
BOT	Build-operate-transfer
C	
CREMA	Contrato de recuperación y mantenimiento (Performance contracts for road rehabilitation and maintenance)
D	
DBFO	Design-build-finance-operate
DBFO/M	Design-build-finance-operate/maintain
DBM	Design-build-maintain
DBO/DBOM	Design-build-operate/design-build-operate-maintain
D-B-B	Design-bid-build
D-B	Design-build
E	
EU	European Union
F	
FBV	Funktionsbauvertrag (Functional construction contracts)
G	
GMP	Guaranteed maximum price
H	
HM	Her Majesty (United Kingdom)

I	
IMF	International Monetary Fund
IRI	International Rough Index
K	
KPI	Key performance indicator
N	
NCHRP	National Cooperative Highway Research Program
O	
OECD	Organisation for Economic Co-operation and Development
P	
PBC	Performance-based contract
PFI	Private finance initiative
PMMR	Performance-based contract for road maintenance and management
PPP	Public-private partnership
PSC	Public sector comparator
S	
SLA	Service-level agreement
SPC	Special purpose company
SPV	Special purpose vehicle
V	
VfM	Value for money
VOB	Vergabe- und Vertragsordnung für Bauleistungen (Procurement and construction contract procedures)
Z	
ZTV	Zusätzliche Technische Vertragsbedingungen
ZTV-MtZEB-StB 01	Zusätzliche Technische Vertragsbedingungen zur messtechnischen Zustandserfassung und Bewertung mit schnellfahrenden Messsystemen (Additional technical specifications for metrological state condition assessment and evaluation through high-speed measurement system)
ZTV-StB 01	Zusätzliche Technische Vertragsbedingungen und Richtlinien für Funktionsbauverträge im Strassenbau (Additional technical specifications and guidelines for functional construction contracts in road construction)

List of Figures

List of Tables

About the Book

Nowadays the public and private sectors are working together for the development of the public infrastructure. Governments have the responsibility to develop infrastructure for social welfare and economic growth. However, the governments of both developing and developed nations often face budget constraints. The public sector, therefore, joins hands with the private sector to overcome budget constraints using private funds (for example, through PPP contracts), or to reduce costs and improve quality using the know-how of the private sector. There are various types of project delivery methods in practice. The most commonly used project delivery methods are design-build, design-bid-build, and construction management at risk.

These days various performance-based contracts are developed and implemented for improved project delivery. The book presents public private partnership (PPP), performance based road management and maintenance contract (PMMR) and Funktionsbauvertrag (FBV) as widely used performance based contracts in infrastructure development. Zimmermann, Josef[1] and Loulakis, Michael C.,[2] among others, define project delivery as the comprehensive process of designing and constructing the physical entity for the owner. The three project delivery methods mentioned above fit this definition. These performance-based contracts expand the definition of project delivery systems and include design, construction, finance, maintenance, and operation of the project.

Performance-based contracts set metrics for the output required of the contractor, rather than the procedures, techniques, and material requirements. The contractor has freedom to be innovative in selecting methods and technology to cut costs and improve quality. PMMR, PPP, and FBV fall under performance-based contracts. From practice, it has been seen that performance-based contracts can cut costs and improve quality. This is one of the major reasons why they are growing in

[1]Project Delivery Systems, Lecture note in Lehrstuhl für Bauprozessmanagement und Immobilienentwicklung an der Technischen Universität München, issue 01/2009. p. 6–20.

[2]Design Build For Public Sector, 2003 Aspen Publishers, Inc, New York. p. 106.

Fig. 1 Concept of PMMR, FBV, and PPP

popularity. PPP has seen increasing popularity worldwide since the 1990s.[3] PMMR is very popular in Latin America and this trend is now spreading all over the world.[4] FBV was developed in Germany and the results have been positive so far on projects implemented under FBV.[5]

This book focuses on the comparative study between these three types of performance-based contracts. The analytical study shows that PMMR, PPP, and FBV are similar in principle (output-based contracts) and contracting period (15–30 years). However, PMMR and FBV are closer to each other in nature, procedure, and implementation than to PPP. PMMR and FBV include only road projects while PPP includes a wide variety of infrastructure. PMMR and FBV do not include financing and operating aspects, whereas PPP does.

As seen in the Fig. 1, PMMR and FBV seem very similar when compared to PPP. The noticeable difference between PMMR and FBV is that FBV includes construction of new roads, while PMMR does not. The other difference is that with PMMR, the contractor is responsible for the management of the road, while it is the duty of

[3] Asian Development Bank. Facilitating Public-Private Partnership for accelerated infrastructure development in India. Workshop report, India, December 2006. p. 19.

[4] Stankevich N, Qureshi N, Queiroz C. Performance-based contracting for preservation and improvement of road assets. Washington, DC: The World Bank; September 2005 (updated August 2009). p. 2.

[5] StrassenNRW Pressinformation. www.strassen.nrw.de/service/presse/meldungen/2008/080703-01.html. Accessed 04 Apr 2010.

the owner in case of FBV. PPP includes financing and operating aspects that are not included in PMMR and FBV. With PMMR and FBV the financing and payments are processed through public funds, whereas PPP projects are financed through private funds, which are reimbursed either by public funds or by direct collection from the users.

All these three contracts follow the same principle regarding risk allocation. The risks are allocated to the partner who can best manage them.

Practical examples of FBV (Road Project: Rastplatz "Blauer Stein"– Miel on the Highway BAB A 61) and PMMR (PMMR in Argentina) have been presented as case studies to help understand these contracts in context of real-world applications. These case studies support the features and characteristics presented in the analytical study.

Abstract

Mobility is the basic need for economic growth and social welfare. Without sufficient infrastructure, especially transport infrastructure, efficient mobility cannot take place. Therefore, the governments of developing and developed countries are investing big amount of its fund in developing infrastructure. The focus of this book is to compare the available performance-based road project delivery systems, which are in implementation worldwide. The book recognizes performance-based road management and maintenance contract (PMMR), Funktionsbauvertrag (FBV, functional construction contracts), and PPP as the widely used performance-based road project delivery systems worldwide. Infrastructure development requires big amount and it is time consuming. Therefore, the public sector and the private sector are working together to deliver the infrastructure in an efficient way. The purpose of the book is to compare the above-stated delivery systems to find out the similarities and differences between them. This would be helpful in order to select the particular method in the particular situation.

PMMR, FBV, and PPP are all "output"-based contracts where the required level of performance is to be delivered. These contracts do not define the process, procedures, techniques, and material requirements. This provides the opportunity for the contractor to be innovative in the selection of methods, techniques, and materials. They are long-term contracts, usually 15–30 years. PMMR and PPP are more widely used in Anglo Saxon countries, whereas FBV is developed and implemented in Germany. PPP has gained its popularity all over the world, whereas PMMR is more popular in South America, although it is also in implementation in the UK, Canada, Sweden, Australia, and Chad, among others. FBV has been developed and is in implementation only in Germany. FBV is developing slowly because from 2002 to 2010 only five projects are implemented and three other are under consideration. They all have well-defined set of performance indicators and service-level agreements to be achieved by the contractor. Funktionsbauvertrag is further divided into contract parts A, B, and C, which are not seen with PMMR and PPP. Part A includes the preliminary works prior to the construction or renovation and is like a conventional contract. Parts B and C are described with the functional requirements.

Part B includes construction or renovation of the road, and part C includes maintenance work. So, FBV cannot be seen as a single contract.

Cost reduction with quality improvement is the basic principle of each delivery method, but not all methods achieve it. The financing with PPP is through the private fund which is later on paid by the government or collected through the users on successful completion. Financial institutions normally allocate higher interest rates on private borrowing than on public borrowing in Germany as well as in other countries. Therefore, PPP suffers additional costs because of the high interest rate, but PMMR and FBV are financed by the government. In this point of view, PMMR and FBV are cost-effective compared to PPP.

They are obliged to deliver the project at the required level, which is also the basic requirement for payment. Therefore, they all are equally efficient in terms of quality. Risk allocation is a considerable factor. Principally, with all these delivery methods, risks are transferred to the partner who can best manage it. However, they differ slightly in allocating unpredictable costs. For example, with FBV (in FBV, the risk allocation is carried only for contract parts B and C), risk resulting from political change is shared among the owner and the contractor. This risk is managed by the owner with PMMR and PPP. With PMMR, the risk allocation for unpredictable costs differs slightly from one country to the other. For example, in Virginia, USA, the contractor bears the risk for unpredictable costs like inflation, accidents, and force majeure, but in Argentina, the costs resulted from force majeure and other disasters are borne by the owner.[1]

PMMR and FBV are similar in nature and procedure compared to PPP. PMMR and FBV are designed specifically for the road project delivery, but PPP delivers other projects like hospitals, prison, and roads. However, principally, FBV could be used for other project delivery like schools, hospitals, and water supply. In this case, functional requirements for all types of projects are to be developed and defined clearly. PMMR doesn't include new construction; therefore, it may be implemented for the renovation and maintenance of other infrastructure other than roads by defining the functional requirements for each project.

PMMR and FBV differ significantly in regard to the implementation area; FBV is designed for new road construction, road renovation/rehabilitation, and maintenance. PMMR is designed for road renovation/rehabilitation, maintenance, and management. PMMR doesn't include new road construction like FBV. However, only one FBV project (A 6 – section between existing point Roth and highway junction Nürnberg-Süd) is considering the new road construction, and all other are considered for road renovation/rehabilitation and maintenance of the road superstructure.

They also differ in the selection of contractor; PMMR selects through competitive public participation with "best value approach,[2]" but the practical examples

---

[1] Stankevich N, Qureshi N, Queiroz C. Performance-based contracting for preservation and improvement of road assets. Washington, DC: The World Bank; September 2005 (updated August 2009). p. 2.

[2] Selection based on the contractor's ability, experience, potential to understand the new approach, ability to handle the risk, knowledge, past performances, qualification of staffs, and technical ability.

show that FBV uses the restricted selection procedure ("nicht offenes Verfahren mit öffentlichem Teilnahmewettbewerb"). The payment and financing with both are through the public funds. Therefore, FBV and PMMR are not identical, but they are similar in principle, procedures, and implications. PMMR has a well-developed structure because it has been implemented since 1998 (Zietlow G, Implementing performance-based road management and maintenance contracts in developing countries – an instrument of German Technical Co-operation, Eschborn, 2004), but FBV was for the first time implemented in 2002, and it is still considered a pilot project. Due to being similar types of project delivery systems, FBV can learn and adopt the features of PMMR. Gerdes, Enno[3] stated that FBV is not a new contracting method; similar contracting methods exist in Argentina, Brazil, Chile, Finland, Guatemala, Canada, New Zealand, Peru, the UK, Uruguay, and the USA. He would have indicated PMMR. This statement also supports that FBV is similar to PMMR.

Two case studies have been included in the book: one on Funktionsbauvertrag (Road Project: Rastplatz "Blauer Stein" – Miel on the Highway BAB A 61) and the other on performance-based road management and maintenance contract (performance-based road management and maintenance contract in Argentina). Case studies presented real-world examples to support the analytical study.

Keywords Infrastructure • Road project • Project delivery systems • Performance-based contract (PBC) • Performance-based road management and maintenance contract (PMMR) • Funktionsbauvertrag (FBV) • Public-private partnership (PPP) • Risk • Payment • Finance • Functional requirements • Owner • Contractor

[3] PPP-Ansätze für Fernstrasse, Dissertation an Universität Hamburg, December 2007, p. 132.

Chapter 1
Introduction

1.1 Background and Basics

Infrastructure is the basic physical and organizational structure (e.g., buildings, roads, power supplies) needed for the operation of a society or enterprise [1]. The infrastructure facilities consist of various economic and social overheads, namely, energy, irrigation, transportation and communication, banking, finance and insurance, science, technology, and other social overheads like education, health, and hygiene (see p. 1 in [2]). Economic development of a country is very much dependent on the availability of country's infrastructural facilities; particularly the agricultural and industrial sectors depend on its infrastructure (see p. 1 in [2]). Infrastructure comprises all the basic needs of the humans; thus, a well-developed infrastructure is the basis for social and economic development. Good infrastructure helps raise productivity and lower the overall costs in the directly productive activities of the economy, but the expansion of the infrastructure should be fast enough to meet the demand for infrastructure in the early stage of the development (see p. 1 in [3]). Therefore, infrastructure is an important factor in the phase of development but is equally important for the developed economies. Without sufficient infrastructure, economies cannot be effective to generate wealth and employment.

Transport infrastructure is one of the infrastructures among others, but it plays a vital role in the economic growth. An efficient transport system is fundamental to the efficiency of any economy; the ability to move materials is an integral part of all sectors – supplies for manufacturing plants, animal feed supplies for farms, retail goods, etc. (see p. 430 in [4]). All modes of transportation are equally important for supporting economic prosperity, the new and improved road networks play an important role in opening up development, but the investment in road infrastructure needs to be part of a long-term policy to promote economic development at national, regional, and local levels. Since all modes are equally important for the economic growth, equal consideration is to be given to the potential of rail, sea, and air transport to achieve economic development (see p. 430 in [4]).

A. Gajurel, *Performance-Based Contracts for Road Projects: Comparative Analysis of Different Types*, DOI 10.1007/978-81-322-1302-4_1, © Springer India 2014

Financing public infrastructure, such as transportation sectors and energy, is enormously expensive, and the construction period is too long (see p. 1 in [3]). Thus, a major question arising in all countries of the world is how to finance the transport infrastructure development. Traditionally, the government has been the source of financing, but the costs of financing the demanded infrastructure are so high that even the developed countries are facing difficulty to fund in the required scale [5]. The developing countries on the other hand are fighting against poverty and are prioritizing health and education sectors; thus, they are facing financial deficits in other infrastructure sector investments, including transport. Because of this situation of permanent lack of funding transportation infrastructure, the alternative ways of financing and project delivery systems have been explored and are widely implemented in the world.

Performance-based contracting has been developed to overcome the disadvantages of the traditional method of contracting and with the objective to help private sectors participate actively with or without private financing. Performance-based contracts can be described "as a goal- or result-oriented process" because it doesn't dictate how to do something except essential needs or minimum functions and standards (see p. 38 [6]). Performance-based contracts reward innovation (innovation in terms of freedom to choose the construction techniques, procedures, and materials) and reward at least the set minimum standards or better; it ensures that public agencies, who are stewards of public funds, make the maximum value of the available fund (see p. 38 in [6]).

Various performance-based contracts have been developed; public-private partnership is regarded as one of them.[1] PPP is "a contractual agreement formed between the public and private sector partners, which allow more private sector participation than is traditional one. The agreements usually involve a government agency contracting with a private company to renovate, construct, operate, maintain, and/or manage a facility or system" (see p. 7 in [7]). The same literature (see p. 7 in [7]) argues that the ownership is retained by the public sector, but the private party is provided with the additional decision rights in determining how the project or tasks are completed. The lecture note of Zimmermann, Josef agrees this definition and states[2] PPP is a "form of cooperation of the public and private sectors for the funding, construction, renovation, management or maintenance of an infrastructure or the provision of a service." PPP is thus a construction contract carried in the mutual understanding and agreements between public and private sectors for the mutual benefits in order to deliver service or facility.

Funktionsbauvertrag is regarded as a performance-based road construction contract which is developed in Germany. This model has been developed mainly in the road and highway sectors by integrating design, construction, and maintenance as

[1] How and why PPP is a performance-based contract will be discussed in Chap. 3.

[2] Paragraph 1, Green Paper on Public-Private Partnerships and Community Law on Public Contracts and Concessions, COM (2004) 327 final, Project Delivery Systems, Lecture note in Lehrstuhl für Bauprozessmanagement und Immobilienentwicklung an der Technischen Universität Müncen, issue 01/2009 pp. 7–28.

the first steps to full life cycle PPPs (see p. 207 in [8]). With PPP road projects, the planning and design are executed by the public authority without the private sector involvement, and the construction is carried out by the private sector. This may lead to higher construction and maintenance costs because of the lack of cooperation on initial phases. With the concept of involving the private sector from the planning and design till the construction and maintenance, Funktionsbauvertrag has been developed for cost efficiency and improved quality.

Performance (output)-based road management and maintenance contract (PMMR) is another performance-based contract in road projects that has been used in Anglo Saxon literature and being implemented in developed as well as in developing countries. PMMR sets the output to be achieved by the contractor not the process or material requirements. PMMR is the road rehabilitation and management contract.

With PPP, the whole financing or a part is allocated from the private sector. But with the performance-based contract for road maintenance and management and Funktionsbauvertrag, the financing is from the public fund.

1.2 Objective and Scope of the Book

It has been recognized that infrastructure is important for development. Initially, the public sector was executing all the responsibilities of developing and maintaining the infrastructure. But nowadays, public sectors are joining hands with private sectors for the development of infrastructure projects. Public funds should cover wide range of sectors, for example, education, health, agriculture, against poverty, and infrastructure; therefore, the governments of developing countries, as well as developed countries, are facing budget constraints in infrastructure financing. Various performance (outcome)-based project delivery methods have been developed either to overcome drawbacks of the traditional delivery methods or to involve private sectors to finance and implement innovative approaches for cost reduction and quality improvement.[3]

This book aims to present the analysis to explore the processes and procedures and all the important aspects that must be present in the Funktionsbauvertrag (FBV), the performance-based road management and maintenance contract (PMMR), and PPP. The book further presents how these contracts are being put into the context of the contracts. The book focuses on the comparative study of these three performance-based contracts depending upon the analysis to explore, what makes FBV and PMMR different from each other and the traditional PPP projects, and how are they implemented and executed.

The scope of the book emphasizes the role of the private sector, financing, risks, as well as contractual terms in the performance-based road management and maintenance contract (PMMR), PPP, and Funktionsbauvertrag. The book will also include a case study on both contracts and interviews with specialists.

[3] These aspects are elaborated in Chaps. 3 and 4.

1.3 Approach and Structure

The structure of the book will present the concepts of various performance-based contracts for the road projects, especially performance-based road management and maintenance contract (PMMR), Funktionsbauvertrag, and an overview of PPP. The book includes their structure, contractual agreements, financing, and a comparative study between them in road projects.

Chapter 1 will deal with a general introduction and background of different forms of performance-based contracts.

Chapter 2 presents the concepts, aspects, and different methods of project delivery systems with the focus on road projects.

Chapter 3 starts with the concept of performance-based contracts followed by the performance-based road management and maintenance contract (PMMR), Funktionsbauvertrag, and PPP for road projects. This chapter will focus on the process,

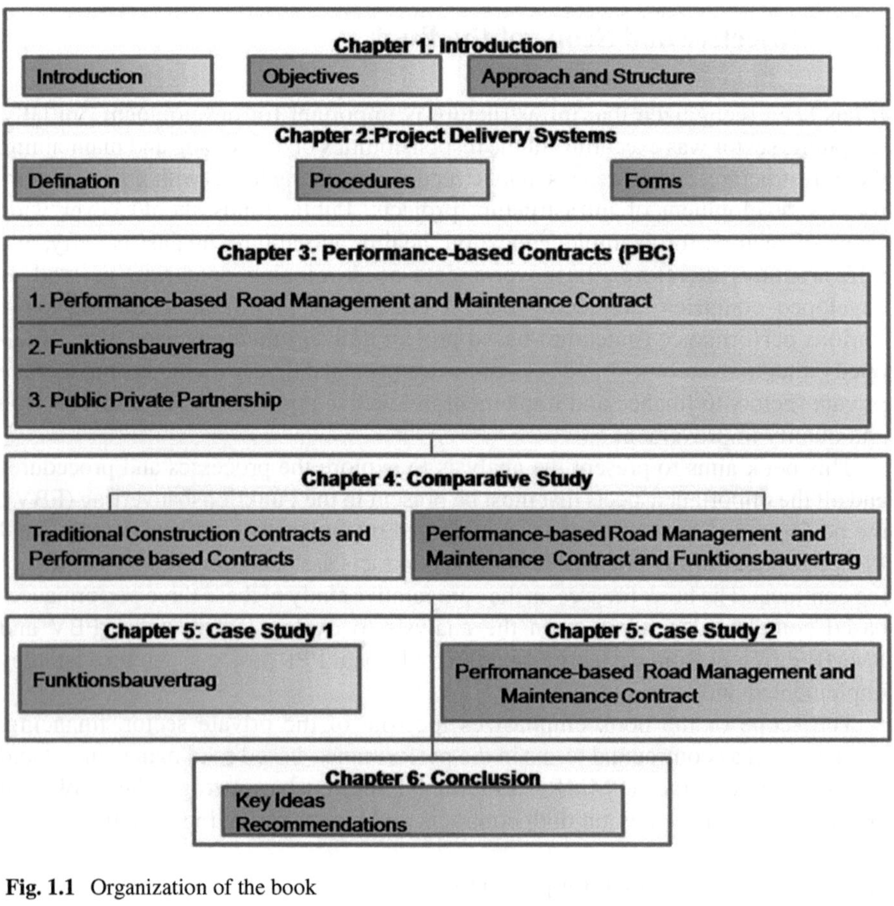

Fig. 1.1 Organization of the book

procedures, the role of private and public sectors, elements of risks, and financing aspects of performance-based road management and maintenance contract (PMMR) and Funktionsbauvertrag. An overview of PPP projects will also be presented.

Chapter 4 focuses particularly on the comparative study of performance-based road management and maintenance contract (PMMR) and Funktionsbauvertrag. This chapter will also present the similarities as well as differences between traditional contracting methods and performance-based contracts.

Chapter 5 proceeds to the case study. The book targets to present a case study on performance-based road management and maintenance contract and Funktionsbauvertrag.

Chapter 6 will conclude with the recommendations that have been gained from the literature review and case study (Fig. 1.1).

References

1. Online Compact Oxford English Dictionary. http://www.askoxford.com/concise_oed/infrastructure?view=uk (2010). Accessed 14 Mar 2010.
2. Kahn NA. Infrastructure for economic development, a comparative study of India and Malaysia. New Delhi: Anmol Publication Pvt. Ltd; 2004.
3. Byoungki KIM. Infrastructure development for the economic development in developing countries, lessons from Korea and Japan, November 2006. Kobe: Kobe University; 2006.
4. Cole S. Applied transport economics, policy, management and decision making. 3rd ed. London: Kogon Page Ltd; 2005.
5. Rodrigue J-P, Comtois C, Slack B. Conclusion: issues and challenges in transport geography. http://people.hofstra.edu/geotrans/eng/conclusion.html (2010). Accessed 14 Mar 2010.
6. Summary of the conference Sept 9–12, 2007, California. Washington, DC: Transportation Research Board; 2008.
7. Jefrey N, Buxbaum IN. NCHRP Synbook 391: Public sector decision making for public-private partnerships. Washington, DC: Transport Research Board; 2009.
8. Alfen HW, Leupold A. Public Private Partnership (PPP) in the German Public Real Estate Sector, Germany real estate – yearbook 2007. The Hague: Europe Real Estate Publisher B.V; 2007.

Chapter 2
Project Delivery Systems (PDS)

2.1 Definition

A project delivery system is the comprehensive process by which a physical entity is designed and constructed. The comprehensive process includes (see pp. 6–20 in [1]):

- "Definition of scope and requirements of a project;
- Procedures, actions, and sequences of events;
- Contractual requirements, obligations, and responsibilities of the parties;
- Interrelationships among the participants;
- Mechanisms for managing time, cost, safety, and quality;
- Forms of agreement and documentation of activity;
- Actual execution of design and construction;
- Closeout of the project and start-up of the new facility."

Loulakis, Michael C. (see p. 106 in [2]) agrees to the above-presented definition of Zimmermann, Josef and states: "Project delivery system is a process or means by which a construction project is comprehensively designed and constructed for an owner, including:

- Project scope definition
- Organization of designers, construction, and various consultants
- Sequencing of design and construction operations
- Execution of design and construction
- Closeout and start up."

A project delivery system defines the way a construction project will be organized so that it is taken from the owner's concept to the physical entity. The organization includes bringing together human and physical resources to accomplish planned objectives (see p. 194 in [3]) with the help of ideas, procedures, personnel, design, construction techniques, schedule and cost estimation, etc., which would lead to the successful completion of the project. This organization should match the owner's design and construction capacities along with the specified requirements of

A. Gajurel, *Performance-Based Contracts for Road Projects: Comparative Analysis of Different Types*, DOI 10.1007/978-81-322-1302-4_2, © Springer India 2014

the project. The specified requirements include the factors like level of quality, time frame, and allocated costs. The organization thus should have the capacity to construct and fulfill these specified requirements. Organization of the project affects the efficiency and effectiveness in designing and building (see p. 35 in [4]).

Every project has its objectives and goals to be achieved; it is thus very important that each involving party should understand the goals, objectives, and obligations. The four prime criteria of the success of project delivery system are costs, quality, time, and safety, although the responsibilities to meet these criteria could be different for different systems (see pp. 6–20 in [1]). Project delivery systems define a common framework for the involved persons and organizations so that each one understands its duties and responsibilities so that they can work within the project in a coordinated manner. The responsibilities, authorities, and relationships are normally defined in the contract documents that provide the blueprint for the project delivery system (see p. 35 in [4]).

Note

The above-presented definition of Zimmermann, Josef and Loulakis, Michael C. considers project delivery systems as a comprehensive process of designing and constructing the project for the owner. Other literatures also denote project delivery system as designing and construction tasks (for instance, Glavinich, Thomas E.: Contractor's Guide to Green Building Construction, New Jersey, USA, 2008). But PPP is also a project delivery method, which includes finance, maintenance, and operation besides design and construction. Therefore, these definitions do not cover PPP. But Sanvido and Konchar [5] define a project delivery system as "the relationships, roles, and responsibilities of parties and the sequence of activities required to provide a facility." This definition could be interpreted in a manner that it includes financing, operation, and maintenance of the project (see p. 3 in [6]). *Thus, for this book, the project delivery systems cover finance, design, construction, operation, management, and maintenance of the project.*

2.2 Phases of Project Delivery

The project delivery starts after the real estate development. Within real estate development, scoping, feasibility study, and programming of the project are carried out. Zimmermann, Josef (see pp. 3–6 in [1]) has also adapted the phase of project delivery according to the Stanford University as presented in Fig. 2.1.

2.2.1 Schematic Design (SD)

It is a crucial phase where expectations are set. Budget and schedule are established; its primary objective is to be determined. The general scope, preliminary design, scale, and relationships among the components of the project, budget, and schedule are clearly defined (see p. 9.1 in [7]). The client specifies specifically about the

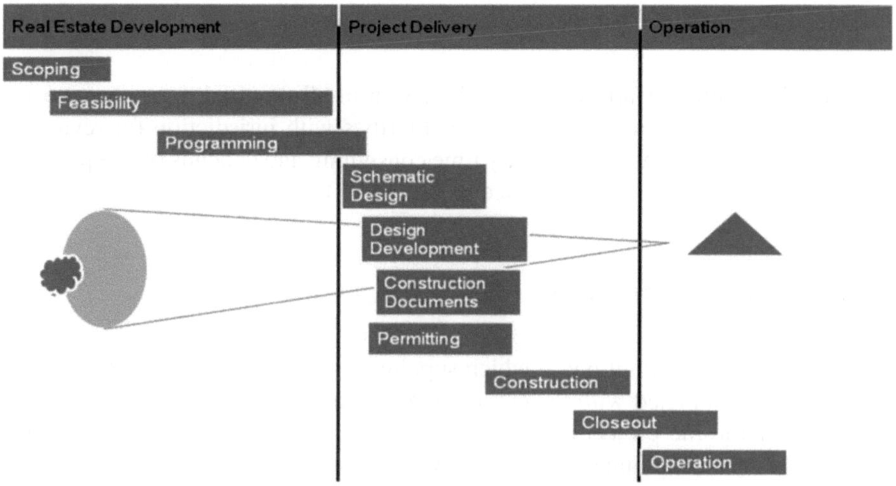

Fig. 2.1 Phases of project delivery (Based on [7]) (financing and maintenance aspects are missing)

requirements for the design option based on the programming phase. This phase should significantly be basic for the technicians and project managers to estimate budget and time schedules. It is regarded as the road map for all involved parties to define the outcomes. Josef Zimmermann (see pp. 3–8 in [1]) agrees to the fact that SD phase involves in setting the expectations, budget, and schedule.

2.2.2 *Design Development (DD)*

Schematic design describes the requirements and scope of the project, while design development refines the scope of previously prescribed in schematic design. The detailed design is developed, and each and every step to achieve the project is defined. Any change in the scope of the project would be possible in this phase because until this phase the participation of user groups is possible, but it should be handled and coordinated properly because it has a big impact on budget and schedule (see p. 10.1 in [7]). Josef Zimmermann (see pp. 3–9 in [1]) also agrees that DD is the last opportunity for any change in scope and design of the project.

2.2.3 *Construction Documents Phase*

This phase concludes the design phase, where drawings and specifications of the project are documented and the contract documents are finalized, which provide comprehensive, fully coordinated construction documents and specifications that the constructor should follow (see p. 11.1 in [7]). Within this phase, all documents required for the approval of the project should be obtained and prepared.

2.2.4 Permitting Phase

The permitting phase starts with the submission of all required documents necessary for the approval from the external authorities with jurisdiction for review of plan and code compliance. It is mostly time consuming and depends on the project's location, complexity, and phasing (see p. 12.1 in [7]).

2.2.5 Construction Phase

This is the most important phase which supplies the need of the client. The aim of the project is completed with the completion of construction. The objective of this phase is to build the project as prescribed in the contract documents meeting the allocated budget and planned time frame. Within the construction phase, the owner, architect, and contractor meet weekly to report and inform about the activities, track submittal status, budget and schedule, and solve the problem raised (see p. 13.1 in [7]). After the completion of construction phase, project closeout begins; in this phase, the constructed project is handed over to the owner for the occupancy.

2.2.6 Closeout Phase

After the completion of construction phase, project closeout begins; in this phase, the constructed project is handed over to the owner for the operation. The goal of this phase is to hand over the finished and inspected project to the owner.[1]

2.2.7 Operation

After the closeout phase, the project is set into operation. PPP road projects are normally operated by the private partner for the contracting period whereby they collect the fee from the users through road tolls. With other project delivery systems, operation of the project is carried by the owner (for instance, with performance-based road management and maintenance contract and Funktionsbauvertrag).

2.2.8 Maintenance

Some project delivery methods include the maintenance work while some do not. Project delivery methods like Funktionsbauvertrag, PPP, and performance-based road management and maintenance contract include maintenance work for the contracting

[1] Based on Stanford University: http://lbre.stanford.edu/sites/all/lbre-shared/files/docs_public/Vol2_1_All-web-1.pdf, p. 14.1.

period, while design-bid-build, for example, does not include maintenance work. These aspects will be elaborated in a later chapter.

2.2.9 *Finance*

Finance is the basic requirement of the project delivery. Project delivery methods like design-build, Funktionsbauvertrag, and PMMR are financed through the public fund. PPP, on the other hand, is normally financed by the private sector and is paid back as part of installments within the contract period (see p. 21 in [1]).

2.3 Key Players of Project Delivery Systems

Within a project, several people, organizations, and entities are involved. The stakeholders of the projects differ according to the nature of the project. According to the definition of project delivery systems, it is the comprehensive activity of the designing and construction of the physical entity. Thus, the key players are the owner, designer, and the constructor. A project, either public or a private, needs permission and final approval from the corresponding authority in order to start the construction work. Therefore, public authority is also regarded as a key player of project delivery systems.

2.3.1 *The Owner*

The client who needs and initiates a project is usually the owner of the project. The owner could be an individual, organization, or any other form of entity. The owner is the entity who holds one or more contracts with the designers and constructors and has the responsibility of essential payment to them. The owner bears all the construction costs. The knowledge of the owner on the design and construction aspects would be very crucial but not the prerequisite. Despite the experience and knowledge, the owner has great expectation from the designer and constructor in the project delivery (see p. 492 in [8]), because the designer and the constructor are hired by the owner as expert to design and build the project.

2.3.2 *Designer/Engineer*

The designer is the expert, who generates a design concept of the project. In case of building construction, an architect is the designer, while in case of road or bridge building, the civil engineer is the designer. The designer is responsible for designing the entity, documenting, and administering the contracts for the construction of the project. The designer generates documents that describe the design intent, and the contractors use these documents to build the projects (see p. 492 in [8]). Design plays a vital role in the quality of the project because the construction is totally dependent on the design. The constructors build the project as prescribed in the design. ASCE Manual No. 73 (see p. 29 in [9]) states that the designer develops project concepts, plans, and design solution to fulfill the project objectives.

2.3.3 *Constructor*

The constructor is responsible to build the proposed project. Constructors can include a variety of subcontractors, suppliers, and fabricators, who design intent of the designer's documents in detail and agree at a prearranged point in the design phase to build the project in a certain amount and certain time frame (see p. 493 in [8]). The constructor should be careful to meet the specified standard as defined in the contract. ASCE Manual No. 73 (see p. 30 in [9]) agrees on the above tasks of constructor and states that the constructor executes the tasks to plan, manage, and properly construct the project according to the plans, specification, and other documents prepared by the designer.

2.3.4 *Public Authorities*

For the public projects, public authorities are the clients and are responsible for the financing, permission, and inspection, but for the private construction, public authorities are responsible for the permission and inspection. The public authority grants permission after inspection on different prescribed aspects such as environment and procedures (see p. 493 in [8]).

2.3.5 *Responsibilities of Different Actors of Project Delivery Systems*

2.3.5.1 Owner

The responsibilities of the owner include the need identification of the project, which is one of the important aspects of any project. Why the project is needed and its goals and objectives should be well identified by the owner. Besides need identification, the owner is responsible to program the project means, he is responsible to make the arrangements of the project, and he sets the criteria and quality level of design (see p. 507 in [8]). Finance arrangement, designing contracts, site allocation, project planning, research (to make the projects successful which includes market research and price research), designing the project to translate the needs and objectives in form of drawings, arrangement of construction contract documents, signing the contract documents, starting the project, and maintenance are other activities that should be carried by the owner (see pp. 6–21 in [1]).

The responsibilities of the owner, however, are not identical for every project delivery system. For example, preliminary designing and maintenance are the tasks of the owner with design-bid-build, but they are performed by the contractor with performance-based contract for road management and maintenance and Funktionsbauvertrag (for details, see Chap. 3). The selection of the appropriate project delivery system is a challenging responsibility of the owner which affects the quality of the project (see p. 23 in [9]).

2.3.5.2 Designer/Engineer

The designer is mainly responsible to present the needs and requirements of the owner in form of drawings or other structures. In order to design, a designer who is normally an architect or an engineer needs to clearly understand why the project is needed and how the needs could be realized. A designer is responsible for programming the project, designing the site and projects, planning the projects' various phases, research (which includes new technology for designing and searching cost-effective method for designing), finalizing the design and administrating the construction contract documents, and giving suggestions during the construction phases (see pp. 6–21 in [1]).

The responsibilities of the designer differ, however, depending upon the types of delivery system (see p. 492 in [8]). For instance, with design-build, the contractor is responsible for designing and building tasks, whereas with design-bid-build, the designer and constructor are different. Therefore, the designer's responsibility ends with the design of the entity in case of design-bid-build.

2.3.5.3 Constructor

The main responsibility of the constructor is to bring the purpose of the project into reality as a physical entity. The constructor is responsible to procure the project, sign construction contract documents, hire the workers, start the construction, inspect during construction to ensure the level of quality, and complete the construction per design and, in some cases, the maintenance works (see pp. 6–21 in [1]).

The responsibilities of contractor differ with the types of the project delivery method. For instance, the contractor is responsible for the maintenance with Funktionsbauvertrag, whereas the maintenance is not the responsibility with design-bid-build.

2.3.5.4 Public Authority

In case of public projects like road construction, the public authority is the owner. In this case, the authority is responsible for the financial burden. The other aspects such as environmental aspects and aesthetical aspects which may be affected by the project are revised by the public authority. If such aspects are under the defined standard values, the authority grants the permission of construction. Inspection is also the responsibility of the authority to ensure if the construction is as per defined (see pp. 6–21 in [1]).

The role of public authority differs with the type of projects. For the private projects, public authority is only responsible for the inspection regarding law, environmental aspects, aesthetical values, etc., and permitting to construct. But with public projects, public authority performs inspection, permitting, and financing tasks.

2.4 Key Factors Affecting Project Delivery Choice

Traditionally, the construction cost was the main criteria affecting the selection of the project delivery method. Normally, the contractor with the lowest bid price was selected, but nowadays, the other variables along with the bid price may affect the choice of project delivery methods. The other variables are construction costs, schedule, quality, risks, and owner capabilities (see p. 493 in [8]). Thus, before deciding on the particular delivery method, various criteria are considered as stated above.

2.4.1 Construction Costs

The owner has the obligation of all project costs. The construction cost is frequently the main concern of design and construction (see p. 493 in [8]). Construction costs depend on the magnitude of the project, but in general construction costs are very high and the owner has limited funds. To meet the defined budget is important, and it is the high priority of each and every member of the project team.

2.4.2 Schedule

"Time is of the essence, is the key factor of the agreement and may be an overriding criterion for completion. The project time available is generally established by the owner in the schedule developed during project conception" (see p. 3.5 in [10]). The delay in the delivery project may add extra costs on the total costs. Mostly the project is defined by the date of completion. The project includes the time frame within which the project has to be completed. Fulfilling the precise schedule would be the most essential consideration in determining how and when a project would be constructed. Delay would add extra costs to the owner. Meeting the schedule is crucial, especially when the interest rate is very high and capital for building is scarce because a small delay raises the costs of construction (see p. 494 in [8]). ASCE Manual No. 73 agrees that owner benefits from completion of a project as soon as possible (see p. 31 in [9]).

2.4.3 Building Quality

"The demand for particular standards of performance in system, finishes, enclosure, or other building elements is directly related to decision about schedule and construction costs" (see p. 494 in [8]). The standard is prescribed in the contracts which should be achieved by the constructor. The designer normally presents a clear level of quantity, budget, and program; thus, change in one of these parameters may lead to the change in others. Sometimes the owner may accept the project with lower quality if the construction cost is lower or the completion of project is before scheduled time frame. Projects with a long life span should assure the quality in order to save maintenance costs (see p. 494 in [8]).

2.4.4 Client Capabilities

The internal capabilities of the client and the client organization could affect the role of client, designer, and contractors significantly (see p. 494 in [8]). Client capabilities and experiences are not the requirements but would certainly affect the efficiency of the construction. The professional expertise and knowledge of the client is assumed as the key to the successful use of the project delivery systems (see p. 3.34 in [10]).

2.4.5 Project Scope

Before the starting of the project, project scope is defined, but the scope is not completely understood or permanently fixed during the course of the project. Scopes are generally defined during design, detail design, and construction phases. The external factors have effect on the scope of delivery system, like change in the price of the material or construction law. The delivery model should to the highest extent recognize the degree to which project could be or couldn't be modified (see p. 494 in [8]).

2.4.6 Risks

Risk is the factor which needs proper management during construction. It should be handled properly to overcome the cost overrun. All project participants should make their best effort to manage and reduce the risks as the project unfolds. The key to reduce risk is to understand the project requirements by all the participants. (Associated risks with the different types of delivery systems are elaborated under respective headings.)

In contrast to Demkin, Joseph A. (Bernstein, Phillip G., FAIA, RIBA, LEED AP), the ASCE Manual No. 73 (see p. 16 in [9]) argues that the project financing has become one of the more significant factors in selecting an appropriate project delivery system.

Different literatures have presented their views on factors that affect the choice of the project delivery method, which are as follows:

Demkin, Joseph A (Bernstein, Phillip G., FAIA, RIBA, LEED AP) (see p. 493 in [8]) in the chapter Project Delivery Methods clearly stated construction costs, schedule, quality, risks, and owner capabilities are the factors that affect the choice of the project delivery method, but it is not clearly presented how these factors lead to the choice of the particular delivery method.

Molenaar, Keith R. and Yakowenko, Gerald (see p. 4 in [11]) argued that choosing the appropriate delivery method requires an adequate understanding of each delivery method, contracting approaches, as well as the ability to understand the advantages and disadvantages of the different delivery methods. They further pointed out the understanding of various contracting approaches and the potential of the management team decisions could be advantageous on deciding the particular delivery method. The selection of the particular method should depend on the project-specific decision processes, on the types of projects, project risks, human resource available, and the objectives of the project.

Ozdemir, Levent (see p. 184 in [12]) argued that there is no single project delivery method that fits in the entire situation and points out the set of goals of the project as a basic for selecting the particular method. The owner's goals can be achieved efficiently if the constructor is motivated, so the focus should be given to that particular method that motivates the contractor to achieve the goals. Factors like the plans and specifications or statements of works, the extent to which the owner is ready to accept the risks, the ability of the owner to manage and monitor the construction works, costs, specified quality, and completion schedule are considered in choosing the particular type of project delivery method. Construction costs, schedule, quality, risks, and owner capabilities are not only the factors affecting the choice of project delivery methods, but financing, project-specific decision processes on the types of project, project risk, human resource available, and the objectives of the project are also the affecting factors.

2.4.7 Conclusion

From the literature analysis, it is difficult to point out the selection criteria of the particular project delivery method. There is not any rule or principle that helps in choosing the particular delivery method. According to Usher, Thomas E. and Davenport, Philip (see p. 58 in [13]), it is difficult to choose the appropriate delivery method for particular project because of the presence of large number of delivery methods. Every delivery method has its unique features, characteristics, procedure, advantages, and disadvantages. Therefore, a good knowledge and comparison of these criteria would be helpful in finding the particular delivery method for the particular projects. Comparison of these criteria with the needs of the project would be another method to choose the particular delivery methods.

2.5 Project Delivery Systems in Road Infrastructure

Sections 2.1, 2.2, 2.3, and 2.4 dealt with the general concept of the project delivery systems. This chapter aims to present the concept of project delivery systems for road infrastructure. "Since 1990, a number of transportation agencies (as owners, sponsors, or contracting agencies of highway projects) have been experimenting with a wide variety of innovative project delivery strategies aimed at lowering the costs and time to produce highway construction and rehabilitation projects, while maintaining or improving project quality" [14]. There are several methods of project delivery like design-build, design-bid-build, PPP, construction management at risks, and Funktionsbauvertrag. Every delivery method is not suitable for every project. For instance, Funktionsbauvertrag and performance-based road management and maintenance contract are delivery methods developed to deliver the road projects; therefore, they are not suitable for building construction or power generation projects with present concepts, procedures, and functional requirements. Funktionsbauvertrag could be used in delivery of other projects including road projects by developing requirements and procedures for each project type. But performance-based road management and maintenance contract does not include construction but includes rehabilitation or renovation and maintenance aspects, so it could be developed for other project delivery including road projects excluding construction aspect. In this case, well-defined requirements and procedures are to be defined for each type of projects.

Project delivery systems is a broad concept compromising all possible delivery methods for all possible types of projects, but project delivery systems in road infrastructure include only the delivery methods that are being used in the road delivery. The road is a public infrastructure and a basic requirement of mobility. Therefore, there is a big investment in the transport infrastructures, particularly with roads all over the world because of its social and economic importance. Various forms of project delivery systems have been widely used all over the world for the construction of roads and highways. Project delivery systems have evolved over time; in the past, the master planner used to design, engineer, and construct the entity. The master builder had control of the design and construction of the project, but the change in technology and increasing complications in construction demanded for the specialization of design and construction (see pp. 6–21 in [1]). Thus, the designer was responsible for designing and the constructor for construction. This idea led to the traditional idea of design-bid-build. With this concept, the communication between designer and constructor is realized to be limited. The communication and interaction occurred only during the end phase of designing and during the construction which resulted in inefficient design, increased errors and disputes, higher costs, and longer schedules. To overcome these disadvantages in the 1970s and 1980s, a third party was brought to assist designers and constructors during designing and constructing. This third party is often called as agency construction managers, but it was not responsible for cost or schedule. In the late 1980s and 1990s, the concept contractor was developed with bounded contracts and was responsible for the work, costs, and schedule; after that, the concept of the construction manager/general contractor or construction manager at risk developed (see pp. 6–22 in [1]).

2.5.1 Forms of Project Delivery Systems in Road Infrastructure

Today, there are choices among various approaches and forms of project delivery systems, in an effort to make construction more efficient (see p. 6 in [15]). Different project delivery systems make use of various kinds of estimation (conceptual, preliminary, and details) at different stages as the project progresses depending on the scope of work (set of expectation of the owner) prescribed at the particular stage (see p. 5 in [16]). Each owner and constructor chooses and implements the particular form of the project delivery system depending on the nature of the project. Whatever form of delivery method is selected, the success depends on the teamwork, how well they are contractually connected to each other, and how well they communicate to accomplish the tasks (see p. 58 in [13]).

Some frequently used forms of road project delivery systems depending upon the practice are described below:

2.5.1.1 Design-Bid-Build

A Comparison of the aspects of project delivery system of Sect. 2.1 with design-bid-build is presented in Table 2.1.

Hence, design-bid-build includes all these aspects of project delivery systems; therefore, it is a project delivery method.

Forms of Design-Bid-Build

Design-bid-build is widely used and well-established project delivery systems (see p. 1 in [17]). The design phase normally accounts for 5–10 % of the total project costs, while the construction accounts remaining 90–95 % of the total project costs (see p. 1 in [17]). Exploratory Study on Responsibility, Liability, and Accountability for Risks in Construction (see p. 20 in [18]) also agrees that in a construction project, design process generally accounts for 5–10 % of the total project costs. The construction phase is costly and time consuming because through this phase the owner realizes his desire in form of physical entity. This data would be very helpful

Table 2.1 Comparison of the aspects of project delivery systems with the aspects of design-bid-build

Aspects of project delivery system	Aspects of design-bid-build
Project scope definition	The owner provides complete plans and specification and set the project for bidding, which is the project scope (see Sect. 2.5.1.1.1)
Organization of designers, constructors, and various consultants	Figures 2.3 and 2.4 present that design-bid-build is an organization of designers, constructors, and various subconsultants. For the design purpose, a designer is hired and a constructor for constructions and consultants for suggestions during the design and construction phase
Sequencing of design and construction operations	With design-bid-build, designer and the constructor are separate entity. Therefore, the owners select the contractor after design phase is completed. The contractor is selected and enters into agreement with the owner to construct the facility as specified.[a] Hence, design and construction are carried out in sequence
Execution of design and construction	Under design-bid-build, the owner contracts separately with a designer and a contractor to execute the design and construction of the project[a]
Closeout and start-up[b]	The owner manages all contracts throughout the construction period up to the completion of the construction and then takes the possession of the project on substantial completion[a]

[a] See pp. 6–26 in [1]

[b]How closeout and start-up are regulated is not described in detail in this book because they are not well described in available literatures. This subject may be interesting for the future research

when the contractor is responsible for both designing and construction tasks and if the payment is on lump sum based. In such case, the constructor knows how much costs to allocate for designing and how much for the construction.

Design-Bid-Build: General Contractor

"Design-Bid-Build is a traditional procurement approach of a project. The contracting agency provides the completed plans and specifications and procures the construction services based on the lowest bid, in sequential order" (see p. 26 in [19]). The design team completes the design, and then the owner contracts the general contractor to build the project depending upon the specified design; with design-bid-build, the general contractor is generally selected depending upon the low price based on the competitive process (see p. 36 in [4]). With this delivery system, the owner deals separately with the designer and the general contractor for the completion of the project. The first step would be to get the designs from the designer in detail. After the completion of the design, the owner typically bids out the project in order to select or open list of qualified contractors (see p. 37 in [4]). After the completion of bidding, the suitable contractor is selected, and the owner and the

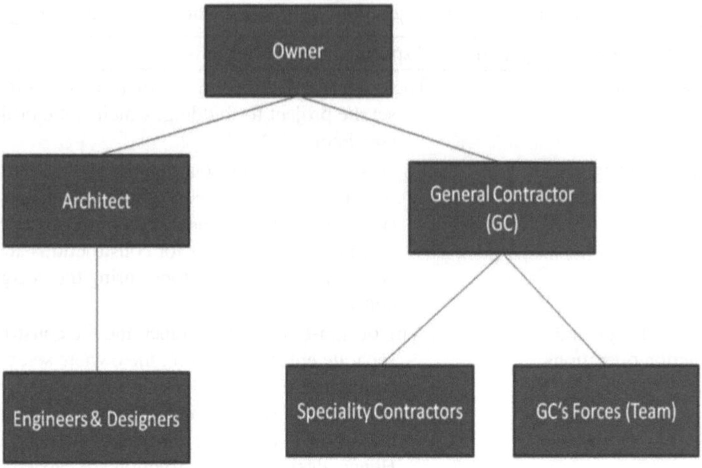

Fig. 2.2 Design-bid-build project delivery system (see p. 37 in [4])

contractor enter into an agreement; the contractor is normally called a general contractor, who is responsible to build the project as specified in the agreement. The general contractor then selects the subcontractors and enters into the contracts for the construction of the project. For example, supplier could be subcontractors.

Design-bid-build has three comprehensive phases:

1. Design: During design phase, the owner prepares detailed project plans and specifications in hiring designers and engineers.
2. Bid: After the completion of designing, the project is set for bidding; the owner invites the construction entity and normally awards to the lowest responsible company.
3. Construction: Building process starts after the selection of the constructors and completion of all contracting documents (Fig. 2.2).

The contractor is not involved during the design phase, which is seen as a disadvantage of this form. Making changes to the design that would reduce the construction costs and wastes and the increase project efficiency after occupancy is difficult to achieve since the owner cannot use the contractor's knowledge in performing value analysis and constructability review in the design phase as well as in cost estimating and scheduling (see p. 37 in [4]).

The general contracts can be a unit price, a lump sum, cost-plus fee agreements, or a guaranteed maximum price (GMP).

Unit Price

Unit price contracts are the contracts where the payment to the contractor is dependent upon the amount of inputs like number of working hours, amount of concrete used, or amount of bitumen used.

Table 2.2 Cost-plus fee contract

Actual costs	Fee as % of costs	Contract amount
1,000,000	50,000	1,050,000

Lump-Sum Contract
With lump-sum contracts, in return for an agreed fixed price as a single payment for the total amount, the contractors provide everything necessary to ensure that the completed project is up to the standard specified and in required level of performance and is completed within the specified time (see p. 1 in [20]). Once the contract is agreed on and signed, the owner is responsible for the payment of the agreed sum regardless of the actual costs beard by the contractors; the constructor would also be paid the agreed price regardless of the actual market price. Thus, the negotiation process mostly takes place before completion of contract documents which would allow discussion among general contractors, designers, and the owner on the issues like safety, schedule, quality, and price (see pp. 6–25 in [1]).

Cost-Plus Fee Agreements
"Under cost-plus-fee contracts, the owner reimburses the contractor for all actual cost associated with the work plus a fixed fee or percentage of cost" (see p. 3 in [21]). Zimmermann, Josef (see pp. 6–25 in [1]) argues plus fee may be a fixed fee, a percentage fee, a multiple of direct costs of labor, materials, and other reimbursable. This type of contract is suitable in the case of uncertainty, when it is difficult to define the scope of the project accurately or the construction needs to be started before full pans or construction needs to be completed than the stipulated time (see pp. 6–25 in [1]).

An example of cost-plus fee contract with 5 % of fee (all costs are presented in $) is shown in Table 2.2.

The contractor receives $1,050,000.

A Guaranteed Maximum Price (GMP)
"GMP is frequently used because it allows owner to gain the protection of the maximum cost of the construction while retaining the potential for saving. It is basically a cost plus fee contract with a cap on it" (see p. 19 in [22]). Jackson, Barbara J (see p. 104 in [21]) argues that it is the agreed sum beyond which the owner is not obliged to compensate the contractor. Thus, the owner is protected by GMP but receives benefit of any realized savings. It allows flexibility in the future prospective and could be negotiated based on the design documents with drawings, costs, time frame, and specifications.

An example of GMP (all costs are presented in $) is shown in Table 2.3.

Design-Bid-Build: Multiple Prime

"Multiple prime is a variation on the traditional approach in which the owner directly retains trade contractors to perform discrete portions of the work" (see p. 9 in [23]). With this, the owner contracts separately with one or more designers and contractors in the project delivery process. Design could be performed by one or more prime

Table 2.3 GMP contract

GMP contract amount	Actual costs plus fee	Contractor impact	Owner impact
1,000,000	1,030,000	Contractor suffers 30,000 loss	No impact. The contract price remains same
1,000,000	999,000	No impact	Owner receives 1,000 savings

Based on Jackson, Barbara J [20], p. 104

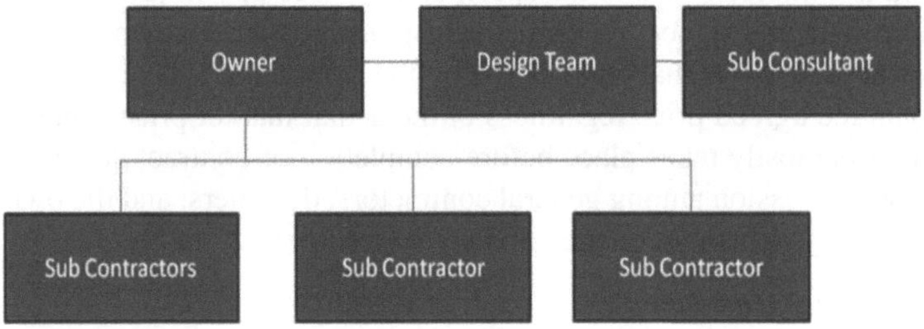

Fig. 2.3 Multiple prime (see pp. 6–23 in [1])

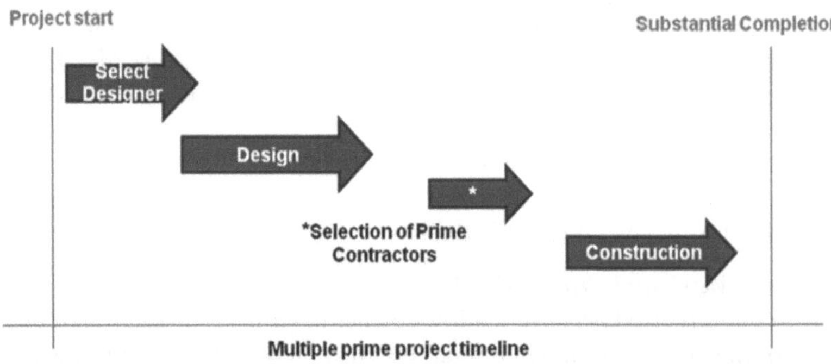

Fig. 2.4 Multiple prime project timeline (see pp. 6–23 in [1])

designers, and the construction is carried through various coordinated but separate prime contracts with multiple constructors (Fig. 2.3; see pp. 6–23 in [1]).

The sequence and the events of the organization are shown in Figs. 2.4 and 2.5. The contractor is normally selected by low bid process. The owner directly contracts with designer and subcontractors.

2.5.1.2 Construction Management at Risk

Comparing the aspects of project delivery system of Sect. 2.1 with the aspects of management at risk, see Table 2.4.

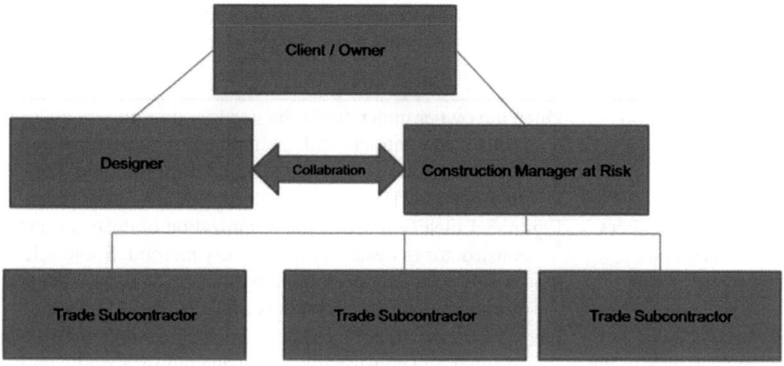

Fig. 2.5 Construction manager at risk (see p. 4 in [24])

Table 2.4 Comparison of the aspects of project delivery systems with the aspects of management at risk

Aspects of project delivery system	Aspects of management at risk
Project scope definition	The owner forms the team comprising designers, constructors, and him to determine the scope of the project before designing and constructing (see p. 52 in [25]) This means project requirements, estimated costs, and time frame of delivery are clearly defined in the contract
Organization of designers, constructors, and various consultants	Figure 2.6 shows the involvement of a designer, constructors, and subcontractors within construction management at risk. The designer involves the specialty subconsultant during the design phase (see pp. 6–26 in [1])
Sequencing of design and construction operations	The owner hires the designer and construction manager simultaneously and completes the designing and then hires subcontractors. The owner then selects a contractor to execute construction management services and construction of the project (see pp. 6–26 in [1]). Thus, this method follows sequential order of design and construction
Execution of design and construction	The designing and construction are executed by the help of the designer and the team of construction managers at risks and other subcontractors (see Fig. 2.6)
Closeout and start-up	After the completion of construction, the inspection phase starts. The owner takes possession of the project after substantial completion (see pp. 6–26 in [1]). The project contracts are also closed with the required signature of owner and contractor

Therefore, the construction manager at risk fulfills the criteria to be project delivery method.

Construction management at risk is a project delivery system where the construction manager is involved from the beginning of the project as the owner's advisor and as the construction manager contracts and coordinates the specialty contractors and subcontractors (see p. 32 in [4]). It is a team-based contract where team building

Table 2.5 Comparison of the aspects of project delivery systems with the aspects of design-build

Aspects of project delivery system	Aspects of design-build
Project scope definition	Once the owner understands the needs and requirements of the project, he contracts with a single entity to perform both design and construction tasks under single design-build contract (see pp. 6–26 in [1])
Organization of designers, constructors, and various consultants	Figure 3.1 clearly presents the organization of designer and constructor in design-build delivery method. Some selected special works are performed by the specialty consultant during design and construction phases (see pp. 6–26 in [1])
Sequencing of design and construction operations	Design and construction are performed by a single entity. Since designer and constructor are carried out by a single entity, sometime they could be executed simultaneously (see below the explanation of design-build)
Execution of design and construction	A single entity carries out design and construction works[b]
Closeout and start-up	After the successful completion of the project, the owner takes possession of the facility (see p. 39 in [4])

among the owner, designer, and the construction manager starts from the beginning of the project's conceptual design till the final construction and the operation of the project. The process is based on teamwork, and every player inputs his/her approaches throughout the project delivery. This method of delivery is flexible in the implementation of late changes in the design process without impacting construction schedules and final delivery dates (see p. 2 in [26]).

The construction manager performs the task according to the specified plan for a fee and reimbursable costs. The construction manager contacts the subcontractors and specialists for the construction procedures after the completion of designing.

The construction manager normally guarantees the maximum price of the project and project time frame to the owner; any saving under guaranteed maximum price could be shared by the owner and contractor or could be on hold by the owner (see pp. 6–26 in [1]). With this system, the owner chooses the designer and construction manager simultaneously to get the design completed, and then the construction manager builds the project with one or more subcontractors. After completion of the project, it is handed over to the owner.

2.5.1.3 Design-Build

Comparing the aspects of project delivery system of Sect. 2.1 with the aspects of design-build, see Table 2.5.

The design-build fulfills the criteria of project delivery systems; therefore, it is a project delivery method.

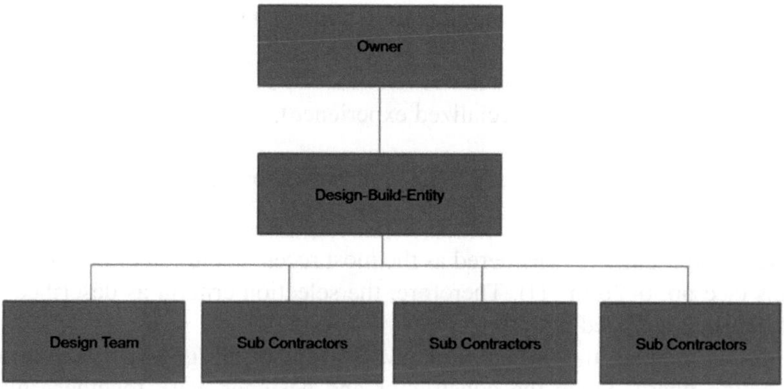

Fig. 2.6 Design-build (see pp. 6–26 in [1])

Design-build is one of the project delivery systems that has been used in road projects as a project delivery system since 1990 with the aim of lowering the costs and time frame [14]. Today, about one-half of all state transportation agencies employ design-build in some form or other in the USA (see p. 110 in [27]). Design-build is a type of contract where a single entity is responsible for designing and constructing a project. With design-build, only one entity or organization is responsible for designing and constructing, and the owner has only one contract for the total delivery of the project; the contract includes both design and construction contracts (see p. 39 in [4]). The design-builder is responsible to design the project based on the performance specifications of the owner, the selection of materials, and methods of construction and coordination of the activities required for the final delivery of the project (see p. 143 in [28]). Although design-build is based on the performance specification, it doesn't fulfill the required criteria to be a performance-based contract.[2]

As shown in Fig. 2.6, the owner needs to administer only one contract for the design and construction. The responsible entity is responsible for both tasks designing and building; thus, the owner needs not to put himself/herself in between the designer and constructors. If there is a failure in the performance criteria, it is the responsibility of the single entity. Whether there is a failure in construction or design, the owner needs not to resolve the issue; rather the single entity is responsible for both failures. The only period that the owner should face responsibility is if there is a change in the agreed-upon project criteria or if the entity suffers with the conditions beyond the control that is not foreseen at the beginning of the project (see p. 40 in [4]).

After the financial arrangement, the owner advertises for the constructors or engineering firm for the entire designing and construction. The selection of the entity is not necessarily in the basic of the lowest bid. The selection is based on the following aspects (see p. 2 in [29]):

[2] See Sect. 3.2.

- "Past performance (both the architectural and or engineering and contractor and major subcontractors);
- Technical approach;
- Technical qualifications (specialized experience);
- Capability to perform;
- Other appropriate factors, such as insurance and bonding; and
- Bid amount."

Note: Design-build is considered as the most recent evolution of project delivery systems (see pp. 6–26 in [1]). Therefore, the selection criteria as described above may have been included with it.

With this method, the project gets involvement of design expertise during the construction phase. Since the constructor and designer work together, this may result in lower costs and lower time frame. The owner has to deal only with the single entity, but the concept of specialization is not applicable here, and the check and balance is also not applicable because of the presence of only one entity for the execution of design and construction tasks.

There are several other project delivery systems that evolved especially for the road project delivery with performance-based criteria like public-private partnership (PPP), performance-based contracts for road maintenance and management, and Funktionsbauvertrag in Germany which would be discussed in detail in the following chapter.

Key Learning of the Chapter[3]

1. Several literatures defined PDS as a comprehensive process of designing and construction of a project for the owner. But it is not true for other PDS like PPP or Funktionsbauvertrag. Therefore, for this book, the project delivery systems cover finance, design, construction, operation, management, and maintenance.
2. Main players of PDS are owner, designer/architect, constructor, and public authority.
3. The phases of PDS are schematic design, design development, construction document, permitting, and construction.
4. Construction costs, schedule, building quality, project scope, risks, and client capabilities are the key factors affecting project delivery choice.
5. There are not any formula, method, and principle that help in finding out the suitable project delivery method.
6. Good knowledge of each project delivery method regarding features, characteristics, procedures, advantages and disadvantages, and the experiences of each player are helpful in selecting appropriate project delivery method.
7. Commonly used project delivery methods in road projects are design-bid-build, design-build, and construction management at risk.

[3] Based on this chapter, the respective references are presented at the end of this chapter.

8. The general contracts can be unit price, lump sum, cost-plus fee agreements, or a guaranteed maximum price (GMP).
9. About one-half of all state transport agencies in the USA employ design-build for road projects.

References

1. Zimmermann J. Project delivery systems, lecture note in Lehrstuhl für Bauprozessmanagement und Immobilienentwicklung an der Technischen Universität München, issue January 2009.
2. Loulakis MC. Design build for public sector. New York: Aspen Publishers, Inc; 2003.
3. Terry GR, Franklin SG. Principles of management. 8th ed. Homewood: RD Irwin Publication; 1987.
4. Glavinich TE. Contractor's guide to green building construction. New Jersey: John Wiley and Sons Publication; 2008.
5. Sanvido VE, Kanchar MD. Project delivery systems, CM at risk, design-build, design-bid-build. Austin: The Construction Industry Institute – The University of Texas at Austin; 1998.
6. Ohrn LG, Roger T. Defining project delivery methods for design, construction and other constructed related service in the United States, Northern Arizona University, Flagstaff.
7. Stanford University: project management, a department of land, building and real estate, Project delivery manual: Vols. 1 and 2. http://lbre.stanford.edu/sites/all/lbre-shared/files/docs_public/Vol2_91_Schematic_Design.pdf
8. Demkin JA (Bernstein PG, FAIA, RIBA, LEED AP). The American Institute of Architects. The architect's handbook of professional practice. 14th ed. Hoboken: Wiley; March 2008.
9. ASCE. Quality in the constructed project: a guide for owners, designers and constructors, ASCE manual no. 73. New York: ASCE; 2000.
10. Construction Specifications Institute. The project resource manual: CSI manual of practice. 5th ed. New York: McGraw-Hill; 2005.
11. Molenaar KR, Yakowenko G (ASCE). Alternative project delivery, procurement, and contracting methods for highways. New York: ASCE; 2007.
12. Ozdemir L. North America tunneling. Lisse: Swets and Zeitlinger B.V Publication; 2002.
13. Usher TE, Davenport P. Fundamental of building contract management, Australia, 2002.
14. Executive Summary, Design-Build Effectiveness Study. Final Report, US Department of Transportation, Federal Highway Administration, January 2006. www.fhwa.dot.gov/reports/designbuild/designbuild.htm. Accessed 20 Mar 2010.
15. Kymell W. Building information modelling, planning and managing construction projects with 4D CAD and simulation. New York: McGraw-Hill; 2008.
16. Pratt D. Fundamentals of construction estimating, delmar. New York: Cengage Learning Publications; 2004.
17. Local agencies and design – build contracting: a briefing paper for legislators and their staffs, Senate Committee on Local Government, Senator Patricia, Chair, State Capitol, Sacramento.
18. The National Research Council. Exploratory study on responsibility, liability, and accountability for risks in construction. Washington, DC: The National Research Council; 1978.
19. Guidebook to highway contracting for innovation: the role of procurement and contracting approaches in facilitating the implementation of research findings, National cooperation highway research program. Washington, DC: National Academy Press; 1999.
20. Institute of Chemical Engineers. Form of contracts, lump sum contracts. 4th ed. Rugby: Institute of Chemical Engineers; 2001.
21. Jackson BJ. Construction management JumpStart. San Francisco: SYBEX; 2004.
22. Levy SM. Construction superintendent's operations manual. New York: McGraw-Hill; 2004.

23. Cushman RF, Loulakis MC. Design-build contracting handbook. New York: Aspen Law & Business; 2001.
24. Cunningham G, AIA. Commissioning large public project using construction manager at risk (CM@R), National conference on building commissioning, New York, 4–6 May 2005.
25. Molenaar KR, Yakowenko G. Alternative project delivery, procurement, and contracting methods for highways. Reston: ASCE; 2007.
26. Kwak ZH, Randall B. Construction management at risk: an innovative project delivery method at stormwater treatment area in the Everglades, Florida. http://home.gwu.edu/~kwak/asce_congress_6.pdf. Accessed 20 Mar 2010.
27. Levy S. Design-build project delivery, managing the building process from proposal through construction. New York: MCGraw-Hill; 2006.
28. Moelmann LR, Harris JT. The law of performance bonds. Chicago: ABA Professional Education; 1999.
29. Satterfield Z. Tech Brief, Publisher by the National Environmental Service Center, West Virginia University, vol. 9, Issue 2, Summer 2009.

Chapter 3
Performance-Based Contracts (PBC)

A performance-based contract is by nature performance based; it sets the performance expected from the final product (project), instead of directing the contractor by the methods to achieve the final project; it means a PBC specifies the desired outcome, not the process to achieve the outcome (see p. 11 in [1]). The standards of the output to be delivered are clearly defined in the project instead of the material to be used of the procedure to be followed. The performance measure should be fair, connected to other systems, and clear and practical. The performance specification differs however from the functional specification as the functional specification specifics the means by which the objectives are to be achieved. The contractor has the freedom during the project delivery in terms of the opportunities to bring innovative approach into the material and process (see p. 22 in [1]). The owner should not be concerned about the project, because in a specified time frame, he or she is delivered with the output at the set standard. The output definition should include desired outcomes for overall contracts, project, or task and the desired outcomes for major task area and should be based on the business result for which the project is constructed and the time of delivery (see p. 67 in [2]). The payment is based on the performance rather than the constructor costs, i.e., PBC pays for the result (see p. 170 in [3]).

The benefits of PBC as mentioned in the literature (see p. 67 in [2]) are as follows:

- Requirements are specified and not the process which enables the contractor to achieve objectives at lowers costs.
- Innovative approach[1] such as selecting the new technology and new methods of construction is encouraged.
- Expectation, output, and accountability are clearly mentioned.

[1] An example of innovative approach in road project: Reduction of construction time by encouraging the sub-contractor by rewarding them on early completion, employing the staffs to simultaneously run the multiple tasks, using high qualified staffs and equipments with high efficiency etc. Reduction of construction time eventually cuts the cost.

A. Gajurel, *Performance-Based Contracts for Road Projects: Comparative Analysis of Different Types*, DOI 10.1007/978-81-322-1302-4_3, © Springer India 2014

- A win-win partnership is established between the contractor and the owner in the sense that they share the risk and reward between them.

The PBC results in the improved outcome performances by addressing three interlocking concepts; these concepts are investments, participation, and incentives (see p. 297 in [4]).

The Technical Report No. 133 of the Federal Facility Council, Washington, DC (see p. 69 in [5]), agrees that PBC has the potential for cost saving, better service, and short delivery times.

Scott, Sidney III and Konrath, Linda (see p. 143 in [6]) agree that the experience and lessons learned from the USA and other countries (e.g., Latin America) have shown that PBC is effective because the contractor has flexibility on selecting input and control over factors that affect performance and is being popular in the USA.

Not all projects are suitable under PBC. PBC has its own characteristics, procedures, and legal obligations (discussed in Sect. 3.3 in detail) which needs to be verified before deciding on PBC. The important frameworks that should be considered are the following (see p. 299 in [4]):

- Evaluation of whether or not the conditions (e.g., if a project is a small one and the owner needs a short-term contract. In this case, the performance-based contract is not the suitable one because PBC by nature is a long-term contract. PBC includes the maintenance aspect. The inclusion of maintenance task makes it a long-term contract) are conductive and suitable to use contracts to drive performance is a very important aspect.
- Whether the set results and outputs (e.g., no ruts for 5 years) could be measured and used to distinguish superior performance from the inferior. Because outputs under PBC should be measurable and comparable to compare them with the specified performance standards.
- Whether the full discussion of the link between performance expectation and the work needed to produce set results can be achieved. Sometimes the performance expected cannot be achieved because of some internal or external factors, such as bad weather condition or available material quality or available workers and equipment. In such cases, it should be analyzed if the quality can be delivered as defined with the help of available resources and conditions.
- Whether the clearly articulated benefits for set performance and consequences for failing the performance could be fixed.

Key Elements of Performance-Based Contracts (PBC)

There are several elements associated with PBC which should be defined in the contracts. The key elements include the following (see p. 16 in [7]):

1. PBC is based on the contract instrument which defines the outcome in terms of results as opposed to methods, procedures, systems, or broad categories of work activities. The principle is what is required rather than how (e.g., technique)

the work is to be accomplished (performance work statement/statement of objectives[2]).

NCHRP Synbook 389 (see p. 4 in [8]) agrees that performance-based contract details the desired outcomes (what is desired), not how, when, or where to do the work.

2. The responsibilities are borne by the contractor for the required performance criteria. The contractor is reimbursed depending upon the set of outcomes which require measurable standards (in meters or liters or km) for the development of the project (right pricing arrangement).
3. The contract needs clearly defined, result-oriented performance criteria and measures. The acceptable level of performance is identified and defined in the contract. In the case of preferable range of performance, the acceptable range must be mentioned. Incentive criteria are created to motivate the constructor to be innovative and to meet or even exceed the level of performance.

3.1 Key Performance Indicators (KPI) and Service-Level Agreement (SLA)

KPIs are a set of measures focusing on the aspects of the project performance that are most critical for the present and future success of the project and are nonfinancial measures (see p. 3 and 5 in [9]). Well-defined objectives and goals are the key for successful road projects and not the performance indicators. The performance indicators are not set only to evaluate the degree to which goals and objectives are achieved but they are used by the road administrator on a particular need to develop or improve performance (see p. 17 in [10]). KPIs are set for the effectiveness of the operation and to achieve or evaluate the set objectives, standards, and performance. The 1997 OECD Report classified the performance indicator according to three prospective (stakeholders): government, road administration, and road user. KPIs can be used in project evaluation, planning, and organization management.

The performance-based delivery method is output based; therefore, KPIs are an important aspect of it. The payment with performance-based contracts depends on the output, not the amount of input (e.g., amount of gravel used) by the contractor. The standard specifications[3] are clearly defined in PBC, which should be met by the contractor. The measurement of the specification is a difficult task to perform; thus, the clear specification measures are specified in the contract. Failure to achieve the defined KPIs may result in contract termination or other penalties (see Table 5.7 for an example of penalty criteria). KPI should be clearly defined in the contract and

[2] Performance work statement/Statement of objectives include the specifications that define the performance requirement of the contract, function specification, performance specification and design specification.

[3] See Sect. 3.2.8 for the example of standard specifications for road projects, which are presented as Key Performance Indicators (KPI).

should be objectively measurable (in units like cm, km, and liters) (see p. 7 in [11]). What performance measure would be applied for what types of application area of the roads and the minimum level of standard to achieve are clearly defined in the contract. Some performance measures criteria are road evenness, grip, minimum level of friction, maximum level of siltation, retro reflexivity of roads, etc. There are no international standard values for KPI regarding road performances; hence, it differs from country to county (see p. 7 and 8 in [11]). (Detail descriptions on the KPIs for road performance-based contracts are discussed in Sect. 3.3.7.)

Service-level agreement (SLA) is an agreement between the contractors and the owner which includes the rules, regulations, and the responsibilities that the contractor and the client should follow until the expiry of the contract. It defines the boundaries in terms of the services that the contractor will deliver to the owner and eventually the user and the volume of work accepted and delivered, plus the metrics to evaluate the performance of the contractor in terms of amount, level, or content (see p. 175 in [12]). It also includes aspects such as definition of service, performance tracking and standard, problem management, owner duties and responsibilities, warranties, legal compliance, and termination of agreement (see p. 23 in [12]).

SLA formally defines the level of service to be rendered and is very important that both parties should understand clearly its term and conditions to avoid future conflicts so that it can result in a long-lasting relationship. SLA defines and identifies the various services expected by the owner and the service level to be delivered by the contractor (see p. 13 in [12]). SLA is established to be able to measure the service level required by the owner and the service level delivered by the contractors.

SLA for road projects includes the following components (see p. 95 in [13]):

(a) Scope and description of the service
(b) Owner service delivery requirements
(c) Costs information
(d) External benchmarking
(e) Performance level and contractor commitment
(f) Process improvement
(g) Issue resolution
(h) Duration of contract (terms)

Note: The level for SLA is defined with the help of KPI (no rutting >20 cm long)

Specific description of the performance level expected is a vital requirement of SLA; the reward criteria for the successful completion and penalties for the unsuccessful completion of the project should be clearly defined in the agreement.

From the definition of PBC, it is clear that it is defined from the prospective of the user; therefore the service level should satisfy his expectations. Payment as well as the acceptance of the project under performance based wholly depends upon the performance level of the result or output. Thus the important aspect of the PBC is defining clear performance standards which should be met by the constructors.

The measures that the owner will use to judge SLA generally fall under the following (see p. 104 in [14]):

(a) Quality
(b) Performance
(c) Delivery/response time
(d) Charges for service
(e) Nature of the interaction with the service provider

The service level on road projects may include aspects such as road safety, road visibility, drainage system, and cleanliness.

KPI and SLA sound very similar but they are not identical. They differ in relation to the scope of the concept; KPI represents essential or critical components of the overall performance, whereas SLA is part of a service contract where the level of service is formally defined (see p. 507 in [15]). SLA defines which service level whereas KPI presents measures. SLA is objectively quantified whereas the KPI is a measure of such quality (see p. 119 in [16]). For example, SLA in a contractual form states that the road should ensure the standard performance level in terms of evenness, grip, friction, availability, and user satisfaction, but KPI states this in measures like vegetation <15 cm and thickness of gravel layer should be 10 cm.

Brown, Paul C. (see p. 228 in [17]) states that SLA consists of KPIs and illustrates an example to differentiate them. He illustrates an example of a hotel service; a hotel team promised its customer to deliver a meal in 30 min; otherwise, it will be for free. This is a SLA. The KPI is the length of exact time it takes to get the meal, in this case 30 min.

For a long time, the performance-based contracts have been accomplished by a set of service-level agreements (see p. 24 in [8]). And the SLA is always prescribed in detail using KPI; therefore, the PBC includes SLA as well as KPI.

3.2 Forms of Performance-Based Contracts for Road Projects

From the literature and practice, the following road project delivery methods could be regarded as the performance-based contracts for the road projects delivery:

1. Performance-based road management and maintenance contract (PMMR)
2. Funktionsbauvertrag (FBV) in Germany
3. Public-private partnership (PPP)

Design-bid-build, design-build, and construction management at risk are not performance-based contracts because of the following reasons:

The delivery methods design-bid-build, design-build, and construction management at risk don't fulfill all the requirements of the PBC to be performance-based contracts as described above under Key Elements of Performance-Based Contracts. With design-bid-build, the owner gets the design completed by the designer; after the completion of design, the owner normally bids out the project and selects the constructor who is responsible to build the project (see p. 37 in [18]). The contracting agency provides the contractor with complete plans and specifications (see p. 26 in [19]), but with PBC, both the responsibilities of designing and constructing are borne by the single entity and only the result is specified, not the process.

With design-build, the first stage involves selecting a design-builder, and in the second phase, the owner works with the selected design-builder to define the project needs and expectation in order to establish measurable performance objectives for the project (see p. 41 in [18]), whereas the performance criteria as well as the result and output are defined before the project is set for bidding in case of PBC. In case of design-build, the owner may provide technical specifications or supply equipment and to that extent become responsible to the contractor (see p.14 in [20]), but with PBC, such criteria are not included. With the design-bid-build, the constructor is normally selected based on low price through a competitive bidding price (see p. 36 in [18]), but this is not always the case with PBC. The selection in PBC is normally based on "the best value" approach (the best value approach is selecting the contractor based on technical and financial ability, past experiences and performances, knowledge, and proposed costs (see p. 5 in [21])) which may not be "the lowest bid" (see p. 3 in [21]).

With construction management at risk, the owner contracts the owner and the designer separately but somewhat simultaneously (pp. 6–26 in [22]). But with PBC such activity is not possible because the designing and the construction are performed by a single contractor. Therefore, it doesn't fulfill all the criteria of a performance-based contract.

Key Learning of the Section[4]

1. PBC sets the performance expected from the final output.
2. PBC is based on the principle "what" is required, not "how" to achieve.
3. Methods, material requirements, and techniques are not defined under PBC.
4. Payment under PBC is fixed price lump sum basis normally through installments (see p. 3 in [21]).
5. Works to be accomplished are defined in performance work statement (PWS)/statement of objectives (SOO) (see Table 3.3).
6. SLA and KPI are established to measure the performance standard of the output under PBC.
7. SLA includes rules, regulations, agreements, responsibilities of each players, and performance level of the project.
8. KPI is the measure of SLA (for instance, the grass along the road side < 10 cm; this is SLA, while 10 cm is KPI).
9. Each PBC is accomplished by SLA and KPI.
10. Performance-based contracts are popular in Latin America and being popular in the USA.

[4] Based on the Sect. 3.1; the respective references are presented at the end of this chapter.

11. D-B-B, D-B, and construction management at risk are not performance-based contract.
12. Performance-based road management and maintenance contract (PMMR), PPP, and Funktionsbauvertrag are widely used performance-based contracts in road projects.

3.3 Performance-Based Road Management and Maintenance Contract (PMMR)

The traditional contract for road maintenance is based on the amount of work measured, and the payment is based on the agreed rates for different work completed (see p. 2 in [11]). In contrast, "Performance based contracting is a type of contract in which payments for the management and maintenance of road assets are explicitly linked to the contractor successfully meeting or exceeding certain clearly defined minimum performance indicators. It is very different from traditional contracting, as only performance indicators are specified, not materials and method for delivery. The government agency must fully define the project objectives, rather than the methods by which they are achieved. Based on the objective, the contractor is asked to determine his delivery methods, equipment, materials, staffing and cost" [23]. (For risk allocation aspect, see Sect. 3.3.3 under Risk Transfer.) The work selection, design, and the selection of the delivery method are all the responsibilities of the constructor (see p. 2 in [11]). In performance-based contracting, the owner does not specify any process or procedures, neither the material requirement, but specifies performance criteria that the constructor is obliged to achieve when delivering maintenance service (see p. 1 in [21]). Based on the objective, the contractor is asked to determine his delivery methods, equipment, materials, staffing, and cost.[5]

Payment in PMMR is made on a fixed price lump sum basis normally through uniform installments, linked to the continuation to meet performance targets. The contractor is not paid for physical work completed, but for the final results (or levels of service) he has delivered through the public fund (see p. 3 in [21]).

According to the World Bank Procurement Guidelines (2004) (see p. 1 in [21]), the performance-based contract is also known as output-based contract which follows the competitive procurement procedures which result in the contractual relationship. "PMMR allocates higher risk to the contractor compared to traditional contract arrangements, but at the same time opens up opportunities to increase his margins where improved efficiencies and effectiveness of design, process, technology or management are able to reduce the cost of achieving the specified performance standards."[6]

[5] http://www.gtkp.com/theme.php?themepgid=118, Accessed on 10.03.2010.

[6] www.performance-based-road-contracts.com, Accessed on 23.03.2010.

The first PBC of road maintenance was piloted in British Columbia, Canada, in 1988 [27]. Performance-based road management and maintenance contract is being used in the upgrading, maintenance, and management of road and highway in countries like the USA, Australia, England, Latin America, the UK, Sweden, Finland, the Netherlands, Norway, France, Estonia, South Africa, Zambia, Chad, and the Philippines, and presently preparations are being made to lunch PBC projects in Albania, Cape Verde, Madagascar, Tanzania, Burkina Faso, India, Cambodia, Thailand, Indonesia, Vietnam, and Yemen [24].

Road maintenance is important because it improves the road conditions which would bring the benefits to the road users through improved access to the hospitals, schools, and markets; improves safety, riding comfort, and speed; and lowers the vehicle operation costs. Negligence of the road maintenance may lead to the need of reconstruction at three times or more the costs, on average of maintenance costs. But many countries spend only 20–50 % of what they should spend for the road maintenance (see p. 1 in [25]). Performance-based contracts for road maintenance is therefore developed for the stable multiyear financing because PBC is a long-term contract, so the government is committed to make multiyear financing for road maintenance (see p. 2 in [21]). This may ensure the proper condition of roads.

A performance-based contract was designed to help the public agencies reap the benefits of private sector innovation [26]. Stankevich, Navaid Qureshi, and Cesar Queiroz (see p. 2 in [21]) agree with this statement and state that the agencies move toward performance-based contracting for road maintenance because it provides incentives to the private sector for innovation and higher productivity.

3.3.1 *Reasons for Implementing (Benefits) Performance-Based Road Management and Maintenance Contract (PMMR)*

The idea of PMMR can be seen as an interest of the government to involve private sector actively with more responsibilities than other contracting methods and for the following reasons:[7]

(a) Reduced maintenance costs are achieved through the implementation of efficient, effective, and innovative techniques, process, procedures, and technology (qualified staffs, efficient machinery, encouraging the staffs by rewarding on early completion, etc.). Since the contract is long term, the contractor implements the quality technique which would eventually cut the maintenance costs. Cost savings in different countries because of the implementation of PMMR are illustrated in Table 3.1.

---

[7] Based on Zietlow, Gunter (see p. 4 in [27]).

Table 3.1 Cost savings of different countries under performance-based contract compared to conventional contract [28]

Country	Cost saving %
Norway	About 20–40 %
Sweden	About 40 %
Finland	About 30–35 %; about 50 % less cost/km
Holland	About 30–40 %
Estonia	20–40 %
England	10 % minimum
Australia	10–40 %
New Zealand	About 20–30 %
USA	10–15 %
Ontario, Canada	About 10 %
Alberta, Canada	About 20 %
British Columbia, Canada	Some, but might be in the order of 10 %

The figures clearly present the reduction of the costs under performance-based contracts for road management and maintenance.

(b) Provide transparency for road users, road administrators, and constructor regarding the conditions of the road. Since the performance is based on defined standard, it should be transparent to all the participants because the road condition should meet the specifications.
(c) The owner has the right to timely control on the specified standards, which eventually improve control and enforce quality standards because if the standards are not met, there is no payment.
(d) Improve overall road condition and road user satisfactions. Performance criteria are based on the user satisfaction. Thus, PMMR improves overall conditions of road.

3.3.2 *Objectives of the Performance-Based Road Management and Maintenance Contract (PMMR)*

It has been realized that performance-based contracting has several advantages compared to traditional contracting; therefore, the road agencies are moving toward implementing the PBC approach over the traditional (see p. 2 in [21]). The key objectives of PMMR could be the following:[8]

1. *Cost savings in managing and maintaining road assets*
   Cost savings with the product quality is the important objective of PMMR. The PMMR can drive to the cost reduction through (see p. 2 in [21]):

   - The use of high efficiency machines, qualified staffs, well-organized plans, etc. from the private sector could be productive to save costs and increase profit.

[8] Section 3.3.2 is based on: Stankevich et al. (see p. 2 in [21]).

- Reductions of administrative and overhead costs are achieved through the better packaging of contracts and lowering the agency personnel for administration and supervision.
- The flexibility of the private sector compared to the public sector in terms of reward performance and punishment in the case of nonperformance.

  The objective of PBC is to improve the road conditions; Australia, New Zealand, and the United States have reported substantial cost reduction through performance-based contracts (see p. 15 in [11]).

2. *Greater expenditure certainty for road agencies*
   The PBC reduces the variation orders and the contractors are paid on lump sum basis normally through uniform installments, throughout the contract period. The risk of costs overrun is the contractor's burden, so the road agency faces fewer unpredictable costs (detail on unpredictable costs is elaborated in Sect. 3.3.3) as well (see p. 2 in [21]).
3. *Ability to manage the road network with fewer agency staff*
   The administration costs with PBC are lower because of fewer contracts between owner and contractor and the quantities of input need not to be measured for the payment procedure. Due to these factors, the road agency manages the road network with fewer staff which ultimately cuts costs.
4. Better customer satisfaction with road service and conditions
   The requirement is based on the outcomes and defined by the user's satisfaction; thus, PBC ensures customer satisfaction. The payment is based on how well the contractor manages to achieve the performance standards defined in the contracts. The needs of the user define the performance standards; thus, better customer satisfaction is ensured by PBC, and regular maintenance over the expiry of the contracts is carried out by the contractor.
5. *Stable multiyear financing*
   PMMR is by nature a long-term contract which covers the period of several years compared to the traditional contracts, and the payment is a lump sum through uniform installments throughout the contract period (see 3.2.3 under duration, and payment and finance). Thus, the government is responsible to make long-term multiyear funding for road maintenance and management.

3.3.3 Features/Characteristics of Performance-Based Road Management and Maintenance Contract (PMMR)

PMMR is a relatively new approach that has been used in the road sector in the past, and today many countries are implementing PBC in road projects (see p. 1 in [21]). PBC in road projects offers many benefits (see Sect. 3.3.1 for Benefits of PMMR), but several features and characteristics are to be considered while implementing it, since the concept is new and the experience with such model is very limited. These features and characteristics are listed below. In order to achieve full benefits of PBC in road management and maintenance, the contractors and the owner need to understand and consider the requirement, risks, advantages, disadvantages, features, process, procedures, and characteristics of PBC.

1. *Risk Transfer*
   More risks are carried by the contractor. In principle, the risks of designing the project like design not matching the specified output and construction are borne solely by the contractor because the contractor designs the project (for road projects there are standards and norms regarding the road design, what should be the color of the road or road geometry etc. are already predefined, so the contractor has little effect in designing). These risks and responsibilities are no more shouldered by the owner in PBC. The contractor is free to make innovative decisions like selection of staffs or material requirements and techniques to perform the tasks and should also bear the risk in the case of failure of his management and innovation and errors in predicting destruction of contracted assets; determining the suitable design, standards, materials, and specifications; in planning the needed maintenance standards; and in determining quantities (see p. 3 in [21]). The constructor is responsible for the technical risks and management risks as well.[9]

   Traffic volume risk (The increase of the traffic volume would make the maintenance work more and eventually raises costs. During the time of contract, the average volume would be estimated. In case of increment of traffic volume, then estimated number would certainly bring more road defects. Such risk is traffic volume risk.) is uncertain and it remains with the client [29]. Uthus, Lillian [30] states that in performance-based contract, the risk caused by the changes in climate and studded tire should be shared between the road owner and the contractor. The other uncertain and unavoidable risks like political risks, environmental risks, and material risks are normally handled by the government. In Virginia, USA, the contractor bears the risks for unpredictable costs like inflation, escalating material prices, accidents, and force majeure activities. In Argentina, the owner reimburses cost overruns if the control is beyond the control of the contractor; these events could be earthquakes or flood or material shortage. In British Columbia, Canada, and Estonia the PBC includes the provision of an annual price adjustment for the changes of price for labor and fuel (Table 3.2; see p. 5 in [21]).
2. *Selection Process*
   Performance-based contract follows the best value approach (depending on knowledge, past performances, experiences, and bid price of contractor) in the selection of the contractor, which must not be the lowest bid approach. With PBC, more responsibilities and risks are borne by the contractor; therefore, the owner must be careful in selecting the contractor, and, therefore, the owner needs to ensure the management capacity, potential, understanding, and ability to handle the projects and associated risks. The selection procedures include choosing a contractor with the capacity to complete the project in time and to choose the right material, suitable methods, techniques, and procedures (see p. 3 in [21]).

[9] Based on Zietlow, Gunter (see p. 3 in [27]).

Table 3.2 Summary of risk allocation with performance-based road management and maintenance contract (Based on above description on heading Risk Transfer)

Risk allocation with performance-based road management and maintenance contract (PMMR)	
Contractor	Design risk, construction/renovation risk
	Technical and management risk
	Performance risk (standard not meeting as predefined)
	Risk that resulted from the failure of the techniques implemented, innovation approach, and
	material used
Owner	Traffic risk
	Political risk
	Environmental
	Material risks (availability)
Sheared	Uncertain and unfavorable risks like unfavorable climate
	Risk by studded (small projection fixed in the tire for better grip) tire

Different countries have different strategies for the compensation of unpredictable costs caused during the contracting period. Factors like inflation, material prices, accidents, and force majeure activities are treated differently in each country (see above examples under risk transfer)

After ensuring the capabilities of the contractor, the selection process takes cost proposal into consideration (see Sect. 3.3.5 for selection procedure).

3. *Payment and Finance*
   Under PMMR, the payment is usually based on the lump-sum fixed price and uniform installments, and the price is paid if the project met the required performance as stated in the contract and is not paid for the input but for the final output or level of service the contractor delivered (see p. 3 in [21]). The payment for the management and maintenance of roads is through the public fund. The payment is made in accordance with the meeting or exceeding defined performance levels of the project
4. *Duration*
   The feature of the performance-based contract is the inclusion of maintenance task. This feature makes PBC a long-term contract because maintenance cannot be carried just after the completion of the project and maintenance has to be carried periodically to ensure defect-free road. The performance-based contract is by nature a long-term contract because the contractor bears more risks and responsibilities and is responsible for the maintenance of deterioration that occurs every year (see p. 3 in [21]). Maintenance is to be carried for a long period which makes the PBC a long-term contract. Normally, performance-based contracts have a duration between 4 and 10 years (see p. 13 in [27]). Stankevich, Navaid Qureshi, and Cesar Queiroz (see p. 2 in [21]) argue that PMMR contracts are usually from 3 to 10 years and could go up to 30 years. The performance standard is revised timely to ensure user satisfaction.
5. *Existence of a Mature and Well-Developed Contracting Industry*
   Performance-based contract is a long-term contract, so the industry with capacity to undertake long-term management, to assume additional and overruns costs,

and to develop necessary programs and quality mechanisms to perform the tasks needs to be selected (see p. 3 in [21]). The selection depends on the experience and past performances of the contracting industry. Therefore, the new evolving companies would not be selected. The mature companies with good past performances are likely to be awarded the PMMR contract.

6. *Private Participation*
   Performance-based contract significantly increases the role of the private sector from the simple execution of works to the management and maintenance of road assets (see p. 20 in [31]). Selecting the work procedures, rehabilitation or renovation, and maintenance are all private sector responsibilities. The private sector performs the tasks of management and choosing the technology for the designing and construction tasks (see p. 1 in [27]). Under traditional contracts, the contractor is responsible for the execution of the work defined in the contract, but with PMMR, the private sector is involved in designing and planning the work to accomplish the project [32], preface page.

3.3.4 Procedures and Approaches of Performance-Based Road Management and Maintenance Contract (PMMR) (Fig. 3.1)

(a) *Establish an Integrated Solution Team*
The first step involves determining the needs, what service or project is needed, and in what level of output needed to be produced to solve the need. The required services or output should be simplified into work orders for the contractors. To satisfy the need, the owner should hire other experts and staffs to build a team.

(b) *Describe the Problems that Needs Solution*
The second step involves defining the problems. Once the problem is defined, the required and preferred solution has to be defined. Defining a problem is the starting point of the project.

(c) *Examine Private and Public Solution*
Once the problem has been recognized, the probable solutions should be considered. Analysis of more solutions presented by different sectors may provide the better option. In such case, both private and public solutions have to be analyzed.

(d) *Develop a Performance Work Statement (PWS) or Statement of Objectives (SOO)*
The next step of the agency is then setting performance requirements and quantity levels; the clearly defined performance work statement (PWS) is then prepared which clearly states the objectives of the project that should be achieved. The poorly formulated PWS affects the output results. A typical PWS/SOO includes the following aspects as illustrated in Table 3.3.

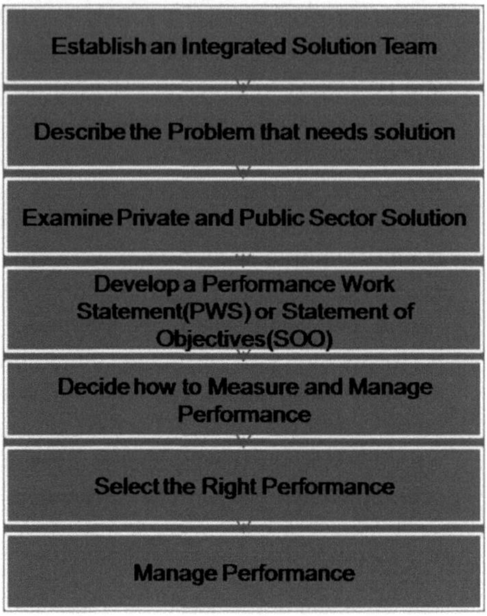

Fig. 3.1 Procedures and approaches of performance-based road management and maintenance contract (PMMR) (see p. 30 in [33])

Table 3.3 Criteria included in PWS or SOO (see p. 42 in [8])

Introduction of the project
Background information
Scope of the project
Applicable documents
Performance documents
Deliverables
Desired outcomes
Required service
Performance standard
AQL (acceptable quality level)
Monitoring method
Incentives and disincentives

(e) *Decide How to Measure and Manage Performance*
The next step is then setting the measures to evaluate the project performance. The performance measures should be defined in the contract to be able to evaluate the project outcomes. It is related to the reward and penalties of the contractors. After defining the PWS and performance measures, the contractor is chosen from competitive bidding. The criteria included in PWS are now defined exactly with the measure (e.g., desirable completion of project is 1 year).

(f) *Select the Right Performance*
Selections of the right performance affect the success of the project. The analysis of similar past projects may be helpful in selecting right performance. The selection of performance indicator with PMMR is based on the road user needs, the expectation of the owner, the level of available fund, etc (see p. 4 in [21]).

(g) *Manage Performance*
The project is then managed and maintained by the contractor based on the required level of performances. The project is then managed and maintained by the contractor for the specified interval of period. And the payment is normally based on the performance or output and paid in lump sum for a series of installments. A typical PWS or SOO defines the following criteria.

(h) *Manage Performances*
The agency needs to define the methodology for measuring the performances. The methods should be simple, practical, and inexpensive. The methodology should be clearly stated in the contract (see p. 4 in [21]). The clearly defined methodologies and measures would be helpful in managing performances and to avoid potential disputes.

3.3.5 The Decision Process in Performance-Based Road Management and Maintenance Contract (PMMR)

Every contract involves series of decision processes for its implementation. PMMR involves mainly two processes according to Stankevich, Natalya, Qureshi, Navaid, and Queriroz, Cesar:[10]

1. Pre-bidding stage
2. Bidding and implementation stage

3.3.5.1 Pre-bidding Stage

Pre-bidding stage engages mainly in the preparation of the requirement of the projects. It is important, powerful, and rewarding to all parties involved including owner, contractor, and end user because the selection of the best contractor brings benefit to all parties involved. It includes all technical details and conditions for optimum safety and reliability before price is fixed and permits technical review before price, and assures the same scope for each supplier (see p. 362 in [34]).

Pre-bidding stage includes the following aspects (see p. 3 in [21]):

(a) *Reasons to consider a PBC approach*
The concept of PBC should be clearly understood by the road agency before considering the PBC. The objective of the project and PBC should be understood and defined in choosing PBC. The objectives could be the following (see p. 3 in [21]):

- To lower the costs
- To implement higher level government directives
- To manage the road network with fewer staff

[10](Section 3.2.6 is based on) Stankevich et al. (see p. 3 and 4 in [21]).

- To receive long-term funding for the maintenance program (see Sect. 3.3.2 under point stable multiyear financing)
- To improve user satisfaction
- To respond to the private sector's offer to deliver more cost-effective maintenance services

On the basis of the main objectives, the road agencies should choose the suitable PBC options.

(b) *Existing legislation*
The selected option or format should match the regulation of the country otherwise the project cannot be realized. The regulatory framework cannot be neglected, so the PBC should follow the legislative framework. If some changes are to be made, the road agency should report to the responsible ministry for amendments on the regulation, but the existing rules must be followed.

(c) *Capacity and changing the role of road agency*
Experience with PBC is very limited and most of the road agencies have no prior experiences. With PBC, the road agency will not be the manager anymore; they would be the owner. This concept needs expertise from the field of transportation and needs new skills (as said earlier, experience with PBC is limited, so there is a need of research work to find new skills in new situation) which should either be developed or experts hired from all over the world (see p. 4 in [21]). Since the procedures with PBC are different than the other types, the road agency should modify the procedures which would fit in the new condition.

(d) *Capacity and unionization of the contracting industry*
The contracting industry should have the capacity in fulfilling the contract. The contracting industry needs to possess the capacity to match PBC requirements. The unionization level of the contracting industries plays a vital role in the success of the PBC. The PBC should be simple or complex or comprehensive, depending upon the capacity of the contracting industries in terms of collaboration. If the contracting industry is small, it should be involve in small and relatively short-term PBC (see p. 4 in [21]).

3.3.5.2 Bidding and Implementation Stage[11]

After the pre-bidding, the project is set for competitive bidding, as discussed in the above chapter; with PBC, the best bidder is selected depending on the best value approach, not the low price. Project implementation is the critical phase because the earlier plan, design, and analysis are to be given in the final shape. Bidding and implementing phases include the following aspects (Fig. 3.2):

(a) *Inventory of potentially contracted assets and determination of their condition*
Before the invitation of the bid, the road agency should clearly define the inventory that needs to be developed and arrange all needed inventory and data.

[11] The chapter is based on: Stankevich et al. (see p. 4 in [21])

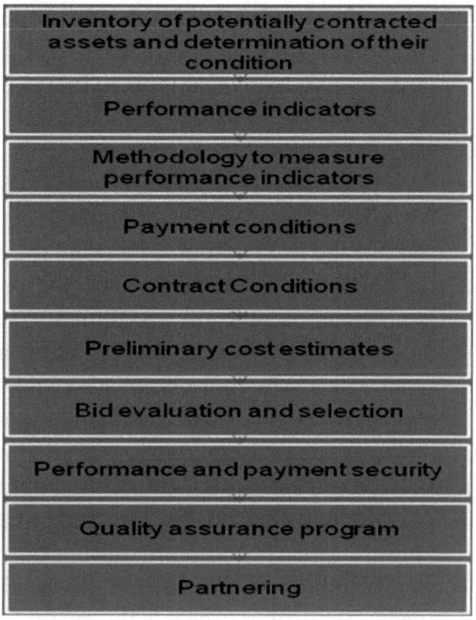

Fig. 3.2 Aspects of bidding and implementing phase of performance-based road management and maintenance contract (PMMR) (see p. 4 in [21])

The condition of the roads needed, the performance criteria, the preliminary cost forecast, and the examining criteria are needed to be specified.

(b) *Performance indicators*
Performance indicators are the measures of the performance. Setting performance indicators is the most important characteristic of PBC. Performance standards should be defined for the project to be developed. The selection and definition of performance indicators include (see p. 4 in [21]):

- Road user needs
- The expectation of the owner
- Affordability on the funding

Achievable and practical factors are determined from the national and international practices and experiences. Is the value 2 for IRI achievable and practical? It is determined from the practices and experiences as well as from the experiments. The long-term performance indicators in case of roads could be the overall condition of pavement, roughness, skid resistance, texture, rutting, surface life, structural condition, etc., and operational indicators include condition of pavement and road furniture (see p. 4 in [21]).

(c) *Methodology to measure performance indicators*
To examine the defined standard, performance measures indicators are established. The agency needs to determine the methods and tools (which tools to be used to measure pavement evenness – either laser measurement or visual inspection) to be applied to measure the performance indicator. These indicators ensure the maintenance and management standards, encourage to work in case

of inadequacies, and promote competitiveness. The acceptance and payment of projects on PBC contracts solely depend on the performance indicators; there are suitable methodologies to be developed to ensure the defined standard (see p. 4 in [21]).

(d) *Payment conditions*
With PBC, the payment is based on the degree that service performed meets the contract standards. The contractor is not paid for the physical work completed but for the final output. The standard of the output should be at least at the level as defined in the contracts and the payment is carried on the fixed price lump sum basic on uniform installments (see p. 3 in [21]). Periodic checking provision should be allocated with the reward and penalty criteria which motivate the contractor to be more innovative and efficient (see Sect. 5.2.6, for the example, of penalty and reward criteria). Zietlow, Gunter (see p. 10 in [27]) states that the performance-based contract for road maintenance and management is basically a fixed price contract. He pointed out an example of a 5-year road management and maintenance contract in Argentina where 55 % of the cost was paid in lump sum in three installments during the first year of rehabilitation work, and the remaining 45 % was paid in 48 equal monthly installments for four remaining years. He further presents the example of New Zealand where the fixed monthly fees are adjusted depending upon the inflation or deflation of the prices.

(e) *Contract conditions*
The responsibilities of the private sector are expanded with PBC; more risk is handled by the contractor. Therefore the contract conditions are needed to be defined clearly with new roles for each partner. The risk is allocated to the party who can best handle them (see p. 4 in [21]) (see Sect. 3.5.7 for risk allocation detail). Contract conditions may include the following requirements (Table 3.4):

(f) *Preliminary cost estimates*
The road agency estimates the total costs of the project that is to be developed. This provides a basis to compare the costs of the various bidders. Preliminary costs provide the idea to the agency to allocate the budget.

(g) *Bid evaluation and selection*
Competitive procurement is applied with PBC. The criteria for selection of contractor under PBC are based on (see p. 4 and 5 in [21]):

- Price only or price and non-price criteria (expertise in the road construction, experiences, past activities, etc.)
- Prequalification of bidders
- Evaluation of technical and cost proposal
- Relevant technical and management criteria
- Past performance, methodology suggested, and technical skill posed

Under PBC, the value-based method is usually applied to select the bidder rather than the low-cost method as discussed in the above chapter.

In Finland, the selection criteria are weighted 75 % to price and 25 % to the technical aspect. In Washington, DC, the selection is based on the technical aspect and the price; the contract price has the biggest weight of 50 %, the

Table 3.4 Example of conditions and performance indicators to be fulfilled in Latin America (see p. 4 in [21])

Culverts and inlets have to be structurally sound and clean to allow for the free flow of water
Surface drainage systems have to be structurally sound and clean to allow for the free flow of water
Vegetation should not exceed a height of 30 cm
No trees should obstruct traffic or pose a safety hazard
Compliance with the program to control erosion
Roadway and right-of-way should be free of litter, debris, and roadkill
There should be no potholes
Cracks more than 3 mm wide should be sealed
Joints have to be sealed
Bridge structures should be clean
Bridge railings should be clean and well painted
Riverbeds have to be clean within 100 m from the edges of bridges
There should be no obstruction of the roadway
Road and traffic signs should be clean
Milestones should be complete, clean, and visible. Missing milestones should be replaced within 24 h
Guardrails have to be clean, complete, and visible
Road markers, road markings, and horizontal road signs have to be clean
Responded in due time to emergencies
There should be no billboards within the right-of-way

Source: Zietlow [27]

technical aspect 20 %, staffing 5 %, management plan 5 %, quality control 5 %, and past performances 15 % quality, but the lowest bidder was not always the winner (see p. 11 in [21]).

(h) *Performance and payment security*
For the long-term contracts, the authority requires performance security and the contractor needs payment security depending on the value of the contract. Performance security requires what if the performance quality is not meet by the contractor? And the payment security requires what if the payment is not done on time or owner is unable to make payment? Under PBC, the bonds could be kept as a security with renewal options or the contracts could be made short term. In Washington, DC, the transport department has introduced the provision of 1-year value bond to solve such possibilities (see p. 5 in [21]).

(i) *Quality assurance program*
Evaluation and inspection are done in a timely manner to check and ensure the performance standard compared to the performance specifications (see Sect. 3.3.7 for performance specifications in detail) in the contracts. The inspection aspects, panel, procedures, and schedule could be defined in the contracts. Normally inspection forum is developed involving representative from the owner, constructor, and supervisor (see p. 5 in [21]). Since the performance is described under user satisfaction, quality should be assured.

Table 3.5 Elements and features of performance-based road management and maintenance contract (PMMR) (see p. 9 in [35])

Elements	Features
Roadway	Potholes, edge raveling, shoving, depression/bumps, joints/cracks, paved shoulders/tumouts
Roadside	Unpaved shoulders, front slopes, sloped pavement, sidewalk, fencing
Traffic service	Pavement markers, stripping, symbols, guardrail, attenuators, signs, object markers/delineators, lighting
Drainage	Side/cross drainage, roadside/medium ditches, outfall ditches, inlets, miscellaneous drainage structure, sweeping
Vegetation/aesthetics	Roadside mowing, slope mowing, landscaping, tree trimming, curb/sidewalk edge, litter removal, turf condition

(j) *Partnering*
Good partnership is required between owner, contractor, and supervisor under PBC as each party has its contribution to the project, and the contract is not about owner's instruction but about road user satisfaction; thus, each party should contribute an establishing working relationship to understand the tasks and risks (see p. 5 in [21]).

3.3.6 Elements and Features of Performance-Based Road Management and Maintenance Contract (PMMR)

Under PMMR, road elements that include roadway, roadside, traffic service, drainage, and vegetation could be maintained and managed. There are several features associated with each element that are to be maintained and managed for smooth traffic flow and safety reasons. The elements and their features associated with roads are presented below. Every feature is specified with the measurable units which are known as key performance indicators, which are described in detail in Sect. 3.3.7. The constructor is required to maintain and manage the road meeting the level of predefined KPI (Table 3.5).

3.3.7 Key Performance Indicators of Performance-Based Road Management and Maintenance Contract (PMMR)

In PMMR, the service-level indicators are the accepted minimum thresholds for the quality levels of the road that should be delivered by the contractor (see p. 8 and 9 in [36]). The performance standards also include the response time and penalties in the case of noncompliance (see p. 3 in [11]).

Performance indicators have to be clearly defined and must be objectively measurable; typical performance indicators for road performances are (see p. 7 in [11]):

Table 3.6 Example of performance indicator applied in different performance contracts in Latin America (see p. 7 and 8 in [11])

Asset class	Component	Performance indicator
Pavement	Potholes	No potholes
	Roughness (asphalt)	IRI < 2.0 (Argentina), IRI < 2.8 (Uruguay)
	Roughness (bituminous)	IRI < 2.9 (Argentina), IRI < 3.4 (Uruguay)
	Rutting treatment	< 12 mm (Argentina), 10 mm (Uruguay, Chile)
	Cracks	Sealed
Gravel surfaces	Potholes	No potholes
	Roughness	IRI < 6 (Uruguay), IRI < 11 (Chile)
	Thickness of gravel layer	10 cm (Chile, Uruguay)
Shoulders	Potholes	No potholes
	Cracks	Sealed
	Joints with pavement	Vertical alignment < 1 cm (Chile, Uruguay), sealed (Peru)
Drainage system	Obstructions	No obstructions. Should allow for free flow of water (Chile, Uruguay)
	Structures	Without damages and deformations (Chile, Peru)
Road signs and markings	Road signs	Complete and clean (Argentina, Chile, Peru)
	Road markings	Complete and visible (Argentina, Chile, Peru)
	Reflectivity of road	160 mcd/lx/m^2 (Argentina)
	Markings	70 mcd/lx/m^2 (Uruguay)
Right of way	Vegetation	< 15 cm height (Argentina, Uruguay)
	Foreign elements	No foreign elements allowed

1. The international Rough Index (IRI) to measure the roughness of the road surface; roughness of the road is related to vehicle operating costs.
2. The absence of potholes and the control of cracks and rutting that are related to the safety aspect and pavement performance.
3. The amount of friction between tires and the road surface that is required for safety reasons.
4. The maximum amount of siltation or other obstruction in the drainage system to avoid destruction of the road.
5. The retro reflexivity of road signs and marking for safety.

Different levels of performance indicators are applied in different countries depending upon the weather, level of traffic, available fund, etc. One example of the different levels of performance indicator applied in performance contracts in Latin America is presented below (Table 3.6).

Key Learning of the Section[12]

1. PMMR is "output"-based contract, which specifies the objective to be achieved, not the process or material requirements.

[12] Based on Sect. 3.3; the respective references are presented at the end of this chapter.

2. PMMR provides freedom to the contractor in selecting process, procedures, technologies, and material requirements.
3. The payment is based on the successful completion of the project, meeting the specified standard. Payment is on fixed lump sum and financing is through the public fund.
4. Normally long-term contract is from 3 to 10 years and can go up to 30 years.
5. PMMR includes KPI and SLA that should be fulfilled by the output; not meeting SLA and KPI may lead to termination of the contract or other penalties.
6. The important KPIs within PMMR include International Rough Index (IRI), potholes, rutting, crack, amount of friction, and reflectivity of roads.
7. PMMR has achieved the reduction of costs in road maintenance and management works with improvement in the quality.
8. The risk of designing and constructing/renovating are transferred to the contractor.

3.4 Funktionsbauvertrag (FBV)

In Funktionsbauvertrag, the quality of the work is no longer described by the materials and construction techniques but the function from the perspective of road a user defined and has been exclusively used in the road's superstructure (see p. 70 in [37]). The result or the output is the important factor in project delivery, and not the way like, methods and procedures. The completed project should meet the performance criteria defined in the contract, and it is the basic requirement for the payment; the payment is carried out after the quality inspection (see p. 227 in [38]). The required performance of the road should be maintained and delivered by the contractor over the expiry of the contract (see p. 70 in [37]). All performance requirements[13] are predefined with the help of measurable functional indicators.

Funktionsbauvertrag is used in Germany as a construction contract in the development of the country's road and highway sectors since 2000. Integrating design, construction, and maintenance are the first steps to full life cycle of PPPs [39]. Funktionsbauvertrag is defined as another PPP model in Germany (see p. 187 in [40]). But it is a contradictory statement because the feature of PPP is financing through the private fund, and with FBV, financing is through the public fund. Operation is carried by the owner with FBV and it is carried by the contractor with PPP. Therefore, FBV cannot be regarded as a form of PPP. It is a civil construction contract in which the private sector doesn't fulfill the public sector's responsibilities; rather it assists the public sector in delivering public service entering the civil construction contracts with the public sector (Fig. 3.3) (see p. 226 in [38]).

The table shows that the FBV includes all the key elements of PBC; therefore, FBV can be regarded as a performance-based contract (PBC).

[13] See Sect. 3.3.5 for the detail on functional indicators.

liters or km etc) for the development of the project (Right pricing arrangement).	measurable standards (like in cm).
The contract needs clearly defined, result oriented, performance criteria and measures. The acceptable level of performance is identified and defined in the contract. In the case of preferable range of performance, the acceptable range must be mentioned. Incentive criteria are created to motivate the constructor to be innovative to meet or even exceed the level of performance	The contract includes clearly defined results, performance criteria and measures. The completed project should meet the performance criteria defined in the contract in order to be accepted from the owner. FBV also provides the oppournities for the contractor to be innovative.

Fig. 3.3 Comparison of elements of PBC with the elements of FBV

Note

The literatures state that the design responsibility is borne by the contractor with FBV. But what the contractor designs is not clear. Regarding road construction and renovation, Germany has its own standard regarding cross section, gradient, drainage system, road geometry, thickness of each layer, color of the superstructure, etc., and all other aspects of road which should be followed by the contractor. So there is no such important role of contractor to design. What is the role of contractor in designing cannot be understood clearly. However the contractor is required to design in detail with clear steps and specifications to achieve the objectives. Therefore, in this book the role of contractor and owner are described as stated in the literatures.

Under Funktionsbauvertrag the acquisition of designing responsibilities is borne by the contractor (see p. 2 in [41]). The contractor performs the design (detail work schedule, applied materials, technologies, schedule, etc.), construction, and maintenance. This provides opportunities for the private sector to apply innovative and cost-efficient approaches from the beginning till the handover phase to achieve specified standards, which would lead to reduced costs. The involvement of the private sector from planning and designing phase till the construction and maintenance phases would bring good coordination; thus, cost reduction could be achieved. Under Funktionsbauvertrag, the financing is not through the private financing but from the public fund (see p. 193 in [40]). Thus, Funktionsbauvertrag doesn't suffer from the high interest rate of financial institute, because the interest rate on private borrowings is higher than the public borrowings. In this aspect, Funktionsbauvertrag achieves overall project cost reduction (Figs. 3.4 and 3.5).

In Funktionsbauvertrag, the contractor is responsible for new construction or renovation or extension and maintenance of a basic road, and the contract is long term for a period of 15–30 years; the contractor is responsible for the maintenance, not for the operation of the road (see p. 8 in [43]). In the year 2002, for the first time in Germany, two road construction projects were awarded as pilot projects under this form of contract (Fig. 3.6; see p. 1 in [45]).

As shown in the Fig. 3.7, Funktionsbauvertrag performs the works related to the road (new construction, renovation, extension, reconstruction, with structural and operational maintenance). The other works related to the roads such as lane marking,

Fig. 3.4 Responsibilities of the contractor in Funktionsbauvertrag (see p. 2 in [41])

Fig. 3.5 Responsibilities of the client in Funktionsbauvertrag (see slide 15 in [42])

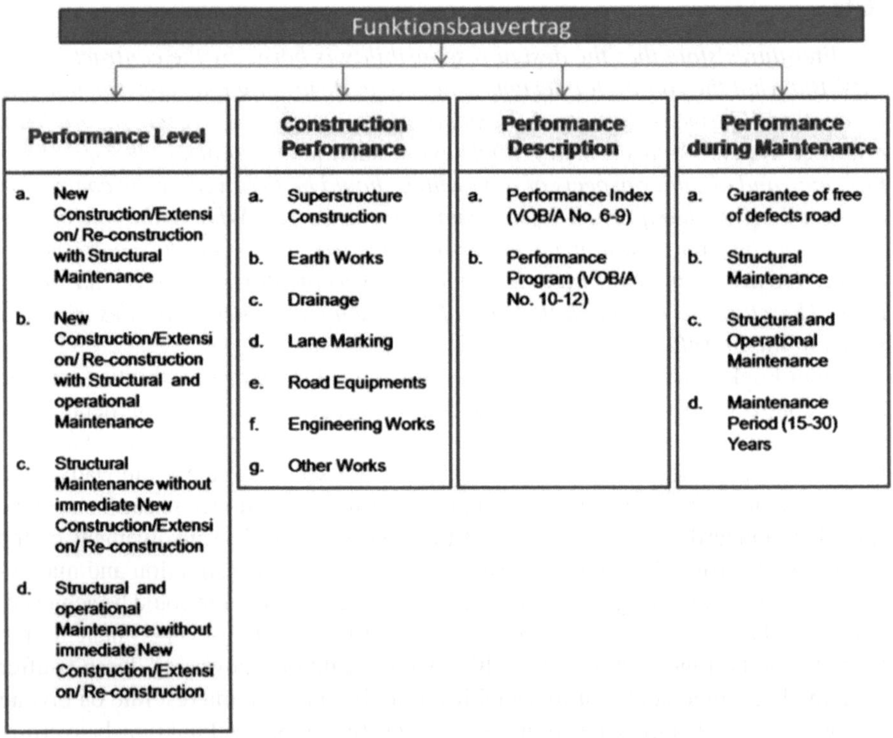

Fig. 3.6 Parameters of Funktionsbauvertrag (Based on Racky, Peter (see slide 8 in [44]))

drainage, earthworks, superstructure construction, traffic-safety-related works, and other engineering works are performed under Funktionsbauvertrag. The performance descriptions are defined by ZTV Funktion-StB (Entwurf), Performance Index (VOB/A No. 6–9), and Performance Program (VOB/A No. 10–12). The maintenance works include structural and operational which should ensure defect-free road for the period of 15–30 years.

Background of the Model		-No separate legal Regulation - In legal Sense as "construction contract" classified
Applied Value Chain		-Construction: New/Renovation -Maintenance
		-Consideration of the A9 project in Thuringia through FBV
Central Source of Financing	Source of Funding	Financing through Public Fund
	Institutional Solution for remuneration Transfer	Financing flow through the public fund to the operator
	Capital Supply	Timely Payment after the completion of the work from the public sector
Geographical Expansion		-Till today project length 10-12 km -Plan with A9 in Thuringia is about 40-50 km to be awarded under FBV
Period of Contract		-The project on A61, A81, A93 with contract period of 20 years, the end of the contract upon reaching ex-ante defined value of equivalent 10-ton axel load transition
		-With the project A31 in NRW, contract period of 30 years

Fig. 3.7 Structure of the Funktionsbauvertrag (see p. 190 in [40])

3.4.1 *Regulations of Funktionsbauvertrag*

According to Nösler, Ingo, (see p. 2 in [41]) in the year 1997, the Federal Ministry of Transport with the help of opinion-building process between the administration and construction management, a research program was conducted; from this research the regulations for Funktionsbauvertrag were developed which is called ZTV Funktion-StB (Entwurf). In this draft, the regulation regarding the performance requirements of the road as well as the required standard measures during the period of the contract is defined; the ZTV Funktion is therefore the part of Funktionsbauvertrag where the following rules are defined (see p. 1 and 2 in [41]):

- Functional requirements of the road surface.
- Target level and standard, values of the substance, and state standard levels for the handover inspection.
- Catalog for the sign of damage.
- Target values and the state values in the maintenance period.
- Also the other rules such as the costs that resulted from the traffic disturbance due to construction.
- ZTV Funktion-StB (Entwurf) stated the rule that the new construction methods would be evaluated through the measuring technique which would compare the state level of the road with prescribed reference value using one of the calculation methods.

The Federal Ministry of Transport also set the following principles for Funktionsbauvertrag (see p. 69 in [37]):

- The construction site should be at least 10 km long.
- Fewer other extra works are to be included.
- Either new route construction or renovation of the existed road surface.

Two property characteristics are applied in determining the road surface properties; they are state characteristics and damage characteristics. The state characteristics are evaluated through the measurement and cover the total road surface where the road length and the cross section are uneven and surface grip and the surface damage are measured, whereas the damage characteristics are visually recorded and in rule locally limited (see p. 69 in [37]). The state characteristics are measured at the time of handover and then every 3 years till the expiry of the contract, and the damage characteristics are at the time of the handover and then every year till the expiry of the contract (see p. 69 in [37]). In the case of the low-standard level than prescribed in ZTV Funktion-StB (Entwurf), the contractor should follow the maintenance measure (see pp. 70 and 71 in [37]).

The implementation of this type of contract is through the state government, but all the applications and operations should be approved by the federal government (see p. 193 in [40]).

3.4.2 Structure of FBV

FBV as defined is a contract for the construction of new roads or renovation of the existing roads and its maintenance for the contracted period meeting the prescribed level of performance without financing and operation from the contractor. The FBV is structured into three contract parts: A, B, and C (see p. 5 in [46]). These parts are performance parts which are divided in relation to the structure of the FBV (KNOLL ET AL (1999, S. 40 f.) und DREHER (2003, S. 260)) (see p. 188 in [40]). These are the contract structures which include the different activities as presented in Table 3.7 (Fig. 3.8).

(a) *Contract Part A* (Traditional Construction Performance):
Structure A includes the works and performance described under VOB/A § 9, No. 6, which are not bound to the superstructure of the road and include only the works which are not defined by the performance level and do not include maintenance prospective (see p. 154 in [48]). It includes works like lane marking, planting trees along roads, earthworks, drainage work, watering the plants, frost protection works, and organization of the construction site.

The reimbursement with the type A is like the conventional construction contract on the unit price basic (see p. 153 in [48]).

(b) *Contract Part B* (Function-Based Construction):
Part B includes the functions and the performances related to the construction and maintenance of the road superstructure; the performance standard is defined according to VOB/A § 9, No. 10 (see p. 153 in [48]). The performance standard

Table 3.7 Contract parts in Funktionsbauvertrag (see p. 3 in [45])

Contract part	Activities
A	Setting construction site
	Traffic flow arrangements
	Earthworks
	Drainage work
	Frost protection work
	Construction work arrangement
	Lane-marking work
B	Construction/renovation of superstructure
C	Functional maintenance of the superstructure

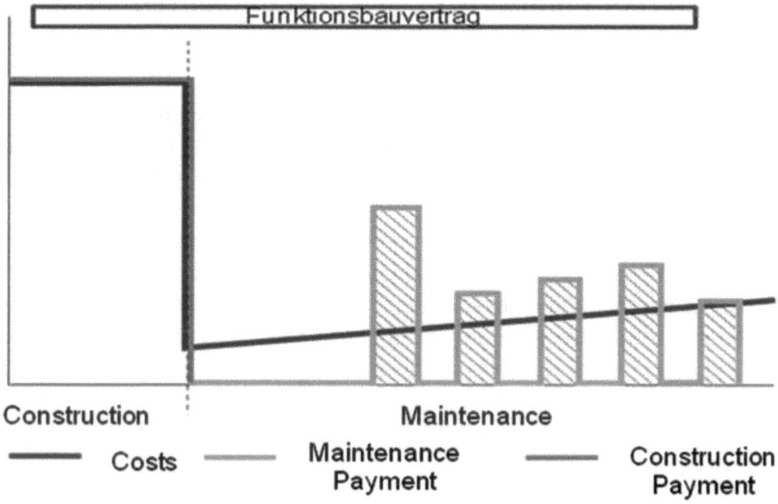

Fig. 3.8 Financing and costs with Funktionsbauvertrag (see p. 9 in [47]). *Note: Explanation of the figure on each contract parts below*

is defined clearly which should be maintained and fulfilled by the contractor as described in VOB/A § 9, No. 10 with standard references.

In part B, the reimbursement is based on the previously agreed amount, and 90 % of the payment is made after the completion of the construction project and the remaining 10 % after the inspection and in case of meeting the defined functional criteria (see p. 189 in [40]). This percentage can differ in practice. With project A 93 under FBV, 50 % of total amount was paid after construction and rest was paid after the inspection [49].

(c) *Contract Part C* (Function-Based Maintenance):
Part C includes all the functions and performances defined in part B; thus, the maintenance and monitoring of the functional performances of the road of part B are carried till the expiry of the contract (see p. 153 in [48]). During the maintenance phase, the functional inspection is conducted every 3 years and the constructor should fulfill the standard at least to the acceptance level defined in the contract.

The payment in part C is made in the annuity basis and the amount is calculated through the present value method (see p. 154 in [48]). The first payment is made in the 9th year of maintenance in the case of positive result of the functional inspection; after that the payment is made in the interval of every 3 years till the contracting period. That means the contractor bears the financing for the first 9 years and then every 3 years of the contracting period for the maintenance of the project; the costs shouldered by the public sector due to the disturbance of the traffic during the construction (because of construction site) are deducted from the reimbursement amount as well as the inflation rate is considered (see p. 6 in [46]).

3.4.3 Features/Characteristics of Funktionsbauvertrag

1. *Risk Transfer*
   The conventional contracts normally include 5 years of a guarantee period. FBV is a new form of contract for construction, developed for the privatization of construction projects which includes the maintenance contract up to 30 years. This is a change from provision to service mentality, which means that whole responsibility and risks are shouldered by the contractor (see p. 37 in [50]). As explained earlier, FBV is output oriented; the contractor is thus obliged to deliver the project with a defined level of performance. It is the whole responsibility of the contractor to deliver the project; this way the owner transfers the maximum risk to the contractor that occurs during design, constructing, and maintaining phases. The tax procedure is not well defined for the FBV by the responsible authority which is seen as a risk for the present context as well as for future prospective (see p. 4 in [45]).

   (a) *Risk Allocation*
   The nature of the payment in FBV results in one particular risk that the contractor is responsible for, the risk associated with the factors that are the determining factors for the payments; therefore, the amplitude of these determining factors determines the amplitude of risks (see p. 132 and 136 in [51]). With FBV, the risks are allocated differently for parts A, B, and C. Part A contract is like conventional contract, so the risk is not transferred to the contractor. Parts B and C are defined with functional requirements; therefore, risks allocation with these parts plays an important role. With part B, all the risks related to design and constructions are transferred to the contractor. The contractor bears the construction and maintenance risks, quality risks, maintenance and construction cost overrun risks and availability risks. If the road is disturbed and not available for the traffic flow, the payment is deducted (see p. 132, 137, 138 in [51]).

   Biller, Nobert (see slide 3 in [52]) agrees with the above statement and states that the contractor bears the risk of design/construction, maintenance, quality, quantity, and prices as shown in Tables 3.8 and 3.9

Table 3.8 Risk allocation with Funktionsbauvertrag (see p. 132 and 137 in [51])

Risks	Contract part A	Contract part B	Contract part C
Construction time/maintenance	No	Partially	Yes
Construction costs	No	Yes	Yes
Quality	No	Partially	Yes
Market	No	No	No
Availability	No	No	Partially

Table 3.9 Risk borne by the contractor in different contract parts of Funktionsbauvertrag (see p. 132 and 137 in [51])

Contract part	Risks
A	No risk is transferred to the contractor
B	Design risk
	Construction risk
	Quality risk
C	Maintenance risk
	Quality risk
	Availability risk

Wiederspahn, Michael (see p. 34 in [53]) argues that with parts B and C, design risks, construction risks, cost overruns because of change in material quantity, maintenance risks, and the risks of other defects on the road are borne by the contractor (Table 3.10).

(b) *Other Risks*
With every FBV, the average traffic volume is estimated (A61 under FBV estimated traffic volume 56,500 per day). If the traffic volume rises, the maintenance work would be more, resulting in additional maintenance costs for the contractor. Such costs are shared among the contractor and the owner. The fluctuation of the interest rate is also considered in the contract. Costs resulting from the inflation and the change of rules and regulations are shared among the owner and the contractor. The compensation amount for such sheared risks is stated in the contract. Force majeure is the risk resulted from the third party intervention or other disasters. Therefore, such risks are handled by the owner. The private part cannot be involved in land acquisition task so the owner himself/herself is engaged in this task and bears the risk arising from it. Since FBV has been explicitly used in the superstructure, the ground construction risk "Grundbaurisiko" is not handled.

2. *Selection Process*
In Germany, FBV is seen as one of the PPP models, which means it would follow the competitive procurement as with PPP, but in practice it's different. For example, with the A 61 project, the selection included the public participants in limited offering procedures ("Nicht offenes Verfahren mit öffentlichem Teilnahmewettbewerb"); there were 21 bidders and 10 were selected for the limited offering procedure (see p. 38 in [50]). The selection procedure is then a

Table 3.10 Other risks associated with Funktionsbauvertrag (Based on Racky, Peter [54], slide 19)

Partner	Associated risks
Owner	Land acquisition risk
	Force majeure
Shared (owner + contractor)	Risks resulted from the change in rules and regulations
	Inflation
	Traffic volume

competitive procedure. Since FBV is a very new method of contracting, limited offering may have been implemented in the start-up to gain experience. FBV is not a concession model rather a pure construction contract, in the completed construction contracts under FBV so long; the procurement process followed a restricted procurement with selected public participation (see p. 132 and 133 in [51]). Ressel, Wolfram [55] states that FBV follows restricted and formalized assessment selection procedures. The restricted procedure ("Nicht offenes Verfahren mit öffentlichem Teilnahmewettbewerb") is a selection procedure where the contracting authority invites a certain limited number of companies for bidding. The invitation includes offering of corresponding benefits to the tender. The competition in this case is limited as the contracting company must be already known by the authority.[14] The competition then starts among the companies, which apply for bidding.

3. *Finance and Payment*
The payment in FBV is through the public budget (see p. 190 in [40]). The private financing concept is not included with it. There are three ways of reimbursement (see p. 189 in [40]):

   (a) Lump sum
   (b) 90% of amount after completion and 10 % after inspection meeting standards
   (c) Annuity payment

   The payment is made after inspection and only if the output meets the predefined performance level (see Sect. 3.4.2 for more information on payment aspect). Large part of the remuneration is done immediately after the completion of the construction or after the evaluation of the handover inspection (see p. 37 in [56]). The remuneration of the maintenance is followed in fixed installments.

4. *Duration*
Part A is like conventional contract so the duration of part A completes as soon as the task is completed. Parts B and C are interconnected and are a long-term construction and maintenance contract in Germany, and it lasts up to 30 years (see p. 37 in [50]). The contractor is responsible for the construction and maintenance for the long duration but is not responsible for the operation (see p. 8 in [43]). During the contracting period, the contractor is also responsible to maintain the project in the standard as defined in the contract.

[14] www.service-bw.de/zfinder-bw-web/generatepdf?type=VB&id...0: Accessed on 11.05.2010.

5. *Private Participation*
With contract parts B and C, design, construction, and maintenance responsibilities are shouldered by the contractor. The contractor performs the planning, design, design and dimensioning, the choice of construction and maintenance strategy, construction, and maintenance; thus, FBV widens the private participation unlike the PPP (see p. 37 in [50]). This definition cannot be true. FBV has been implemented so far only in the road projects. So the corridor specification cannot be carried by the private sector. The road is planned by the government. This means, in FBV, the contractor performs design, design and dimensioning, the choice of construction and maintenance strategy, construction, and maintenance. In principle, the contractor has freedom to design. But the standards for the road (like geometry, thickness, and color of superstructure) are already defined in each country. This means the contractor has little freedom in designing.

3.4.4 Implementation of Functional Indicators

With conventional contracts, the required quality is through the technical rules and regulations on the construction process, material, and mixture defined. But in FBV, such technical regulations play only a minor role; rather the required functionality is the major objective to be achieved and determined by the state condition through measurement and visual inspection [57]. Thus, functions of the road and user satisfaction are the main targets of FBV. The functional standard of the road is regulated through the state condition and damage condition of the road; the standard values to be met are predefined in ZTV Funktion-StB (Entwurf). To be specific, for the FBV, the additional technical requirements are defined; they are ZTV Asphalt or ZTV Concrete and presented by ZTV Funktion-StB 01 (additional technical specifications and guidelines for Funktionsbauverträge in road construction) and ZTV-MtZEB-Stb 01 (additional technical specifications for metrological state condition assessment and evaluation through high speed measurement system) (see p. 69 in [37]). The first two pilot projects under FBV are regulated through ZTV Funktion-StB 01 and ZTV-MtZEB-Stb 01 (see p. 6 in [58]). The contract includes which methods are to be applied (asphalt or concrete) and the standard conditions for either pavement. Damage characteristics are described by the possible damage that can occur during operation, and the state condition is defined by the user's satisfaction (see p. 6 in [58]).

3.4.4.1 Damage Characteristics

The damage characteristics are recorded visually for the inspection. The damage characteristics are categorized under different conditions and aspects of the road. The damage aspects that are listed under ZTV Funktion-StB 01 are presented in Fig. 3.9.

Assessment will be based on the aspects listed in Fig. 3.12 based on the annex of the ZTV Funktion-StB 01 (see p. 8 in [58]). As shown in Fig. 3.10, the

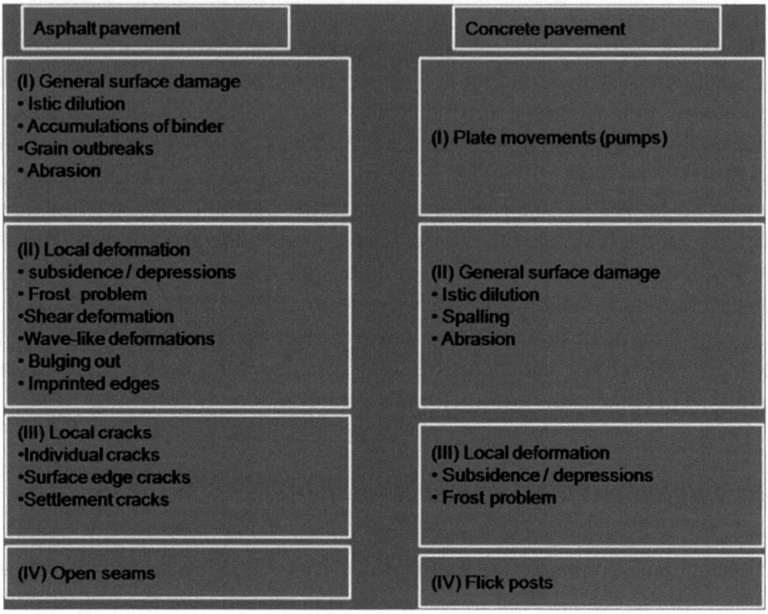

Fig. 3.9 Damage characteristics as defined by ZTV Funktion-StB 01 (see p. 5 and 6 [59])

damage characteristics are presented with the scale from 0 to 3 during inspection (see p. 8 in [58]):

0 – Without Damage
1 – Slightly damage
2 – Medium damage
3 – Severe damage

Guidance on the damage criteria

Figure 3.11 presents an example for the damage characteristics in case of binding material enrichment.

3.4.4.2 State Condition

State condition is defined based upon the functionality of the road. The road should ensure the user's satisfaction. The functional requirement is the basic aspect of FBV that should be maintained by the contractor. ZTV Funktion-StB 01 defines the aspects that fall under state condition of the road which are presented in Fig. 3.12.

The aspects of the roads presented in Fig. 3.13 are the factors for determining the functionality of the road. For every factor, a minimum requirement value is

Asphalt binding (stability)		
I	General superstructure damage	(Ib) Binding material enrichment

Damage description:

The binding material is seen on the surface of the road, the overflow (overgreasing) is clearly seen on the warm weather.

Damage statement and –severe:

-Eyes recognisible features
-The assessment of the features are carried in terms of quality:
(0) without damage:
(1) Little damage: visible darkening due to high coherent binding material
(2) Medium damage: cleatly visible of the darkening in the tyre tracks due to high coherent and free binding material
(3) Severe damage: visible wet surface in the superstructure, specially in the tyre tracks, tyre niuse during warm weather in the rain wet road

Fig. 3.10 Damage characteristics – asphalt pavement (see p. 2 in [59])

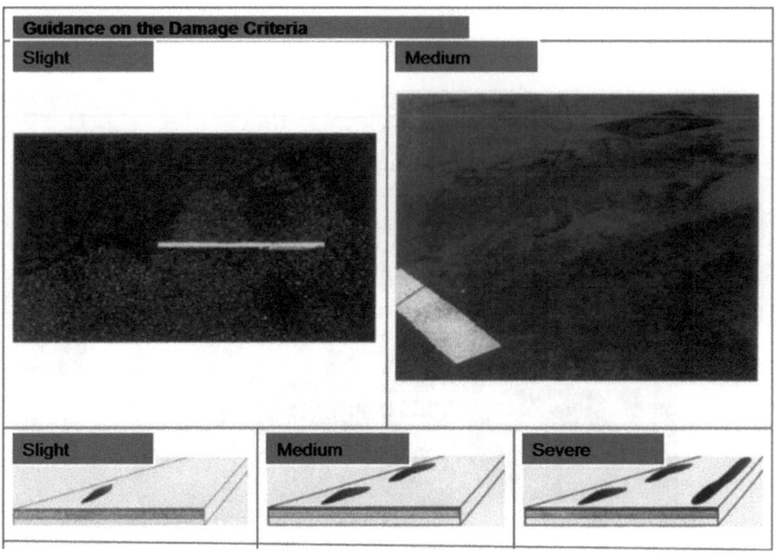

Fig. 3.11 Damage criteria – binding material enrichment (see p. 2 in [59])

defined by ZTV Funktion-StB 01, which should be fulfilled by the contractor. The measurement technique is prescribed by the ZTV Funktion-StB 01. The state characteristics are measured physically with the scale 1 (very good) to 5 (very bad), and then the determined state vale of the road is converted to the standardized value with the help of ZTV Funktion-StB 01-defined standardized function (see p. 9 in [58]).

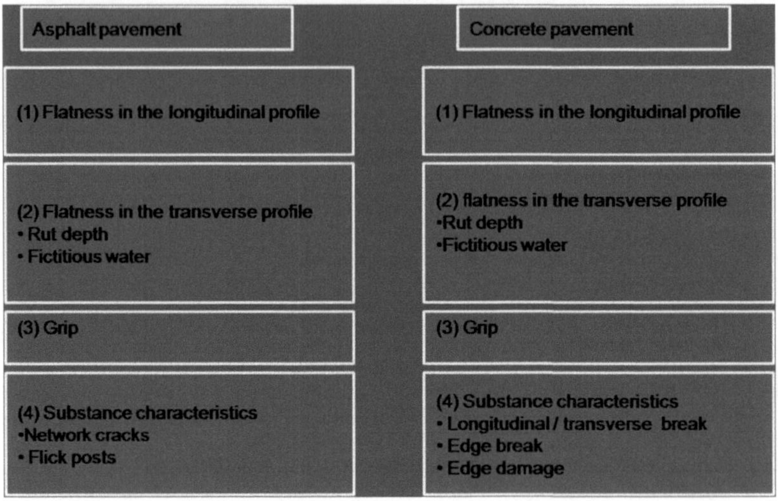

Fig. 3.12 State condition as defined by ZTV Funktion-StB 01 (see p. 4 and 5 in [59])

Fig. 3.13 Grip state condition assessment (see p. 4 in [37])

During the inspection for road state condition, the longitudinal and transversal evenness, the grip condition of road, and superstructure defects (cracks and ruts) are measured. The measurement is carried visually with the help of cameras and through laser measurement as shown in Fig. 3.14.

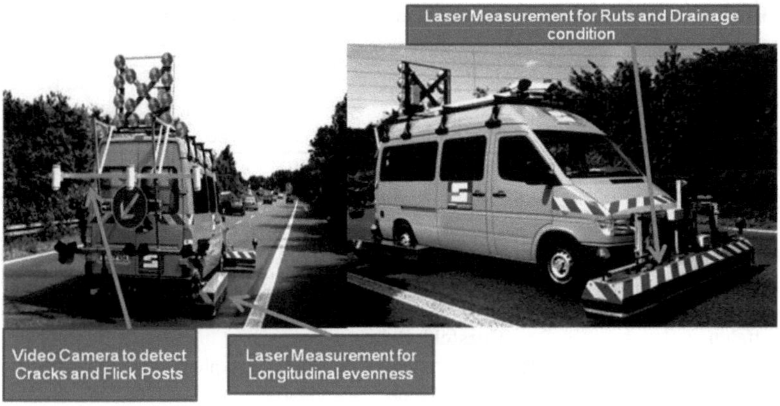

Fig. 3.14 State Condition assessment of the road (see p. 4 in [37])

Process and Requirements for the Maintenance Period

The maintenance is carried in contract part C of FBV. The maintenance period begins with the end of construction and ends with the expiry of the contract. At the end of contracting period, acceptance inspection is conducted by the owner. During the maintenance period, the owner conducts the operation inspection on the rolling basis of every 3 years on state value and damage characteristics to ensure the meeting of the standard functional requirement of the road and on the expiry of the contract period. The acceptance inspection is conducted to ensure the defined state value; on meeting the defined state value, the contractor is no more responsible for the project (KAPPEL (2003, S. 265)) (see p. 188 in [40]). Slight damage could appear during the maintenance period, at the inspection time, and at the handover inspection, but in case of medium or severe damage, the measures should be carried out locally, and the contractor is obliged to maintain the road at predefined standard.[15] For the project acceptance, at least "good structural condition" of the road, i.e., the state condition <= 1.9, is required during the inspection (see slide 7 in [52]).

3.4.5 *Experience with Funktionsbauvertrag*

In year 2002, for the first time, two pilot projects were implemented in Germany in the form of FBV. They are (see p. 260 in [60]):

1. A 81 in Baden-Württemberg: BAB 81 by Rottweil, 10 km Rehabilitation
2. A 61 in Rheinland-Pfalz: BAB A 61, AK Koblenz – AS Kruft, 10 km Rehabilitation

[15] Based on (see p. 16 in [59]).

Table 3.11 Funktionsbauverträge in the Federal Highways ([61] based on BMVBS (2009a, S. 52) and Racky, Peter [42], slide 12)

State	Federal highway (BAB)	Section of highway	Year	Pavements type	Contract duration (years)	Length (km)
Rheinland-Pfalz	A61	Koblenz–Kruft	2002	Concrete	20	10
Baden-Wūrttemberg	A81	Obemdorf–Rottweil	2002	Asphalt	20	10
Bayem	A93	Brannenburg–Kiefersfelden	2003	Concrete	20	11.7
Nordrhein-Westfalen	A31	Gescher/Coesfeld–Reken	2004	Asphalt	30	12.7
Nordrhein-Westfalen	A61	Rastplatz "Blauer Stein"–Miel	2007	Asphalt	15	8.7

Further there have been two more projects the following year under FBV. They are:[16]

1. BAB A 93 between the junction of Brannenburg and Kiefersfelden in Bayern which is 11.7 km long and for rehabilitation and contract period of 20 years.
2. BAB A 31 in Nordrhein-Westfalen between the junction of Lembeck and Geschel/Coesfeld which is 12.7 km long, and for rehabilitation, the contract period is 30 years.

The other projects under Funktionsbauvertrag on BAB A 61 in Nordrhein-Westfalen between Rastplatz "Blauer Stein"– Miel, which is 8.7 km long, are presented in Table 3.11.

There is also consideration in Thuringia to award a FBV for the extension of the road section of A 9 by 40–50 km with the operation service to some extent.[17]

In another project in North Rhine-Westphalia on the motorway, A 59 between Düsseldorf and Monheim tender repealed because the federal government was against the expectations of the state and did not allocate additional budget for this project.[18] In addition, further road projects under FBV are in consideration and planning (see p. 92 in [61]).

Another project is implemented on the BAB A 6 under Funktionsbauvertrag. The project is for the expansion of A 61. It would be the first project under Funktionsbauvertrag which would cover the earthworks, superstructure construction, noise protection works, landscaping, and bridge construction works, because the

[16] According to Werner Bednorz (BMVBW, Referat S 17: Straßenbautechnik und Straßenbeanspruchung) - Beckers, Thorsten (see p. 191 in [40]).

[17] According to Herrn Lutz Irmer (Thüringer Landesministerium für Bau und Verkehr, Abteilung Verkehr, Leiter) – Die Realisierung von Projekten nach dem PPP-Ansatz bei Bundesfernstraßen, Dissertion, TU Berlin, Fakultät Wurtschaft & management, Berlin, 2005, p. 191.

[18] According to Ulrich Habermann (Hauptverband der deutschen Bauindustrie e.V., Abteilung Verkehrswegebau) – Die Realisierung von Projekten nach dem PPP-Ansatz bei Bundesfernstraßen, Dissertion, TU Berlin, Fakultät Wurtschaft & management, Berlin, 2005, p. 191.

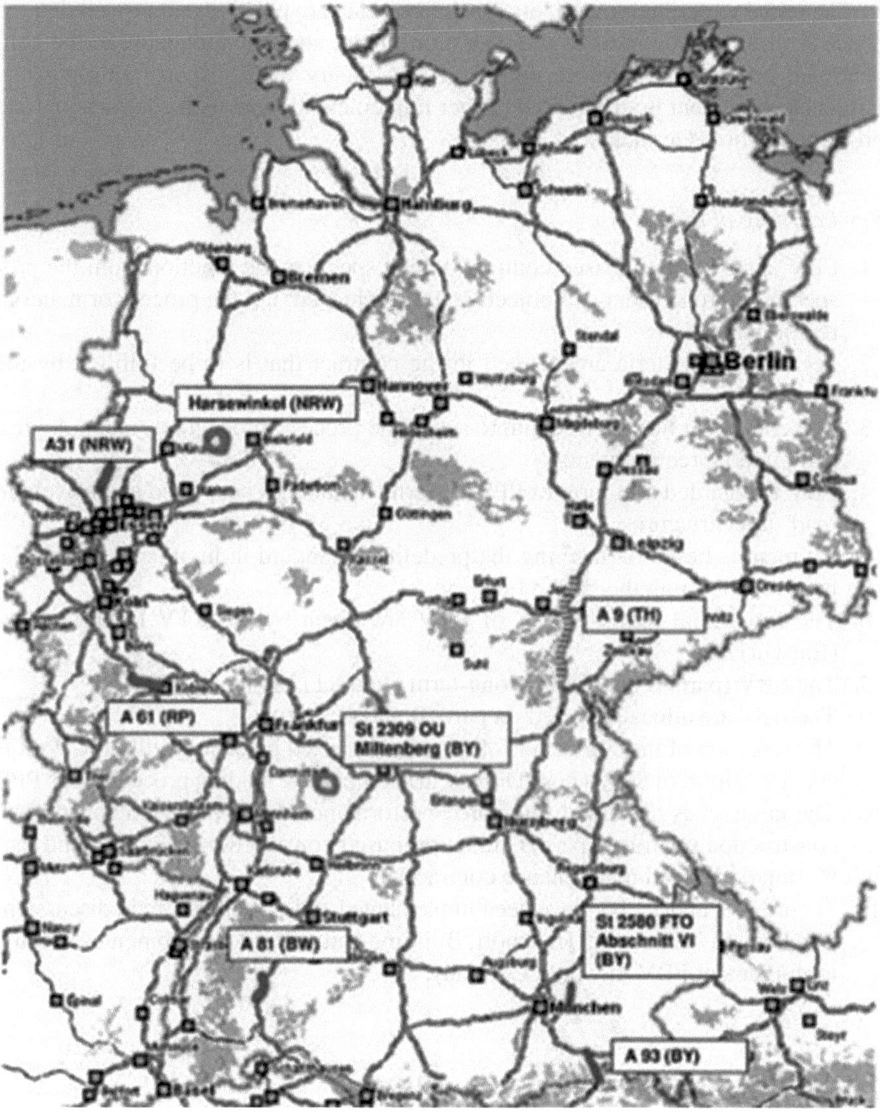

Fig. 3.15 Pilot projects with Funktionsbauvertrag (see p. 228 in [38])

previous projects under Funktionsbauvertrag covered only superstructure works. The construction started in year 2010. The project covers the expansion of highway (A 6) section between existing point Roth and highway junction Nürnberg-Süd. The project cost would be 65 million euro with contract period of 25 years (Fig. 3.15).[19]

[19] Bayerische Staatregierung: www.bavaria.de/Pressemitteilungen-.1255.10242857/index.htm, Accessed on 31.05.2010.

The initial experiences with this type of contract are positive and proved that the concession-based contracts in the operation model are also applicable in the conventional contracts. Therefore, the Federal Ministry of Transport, Building and Urban Development is striving for further implementation of such contracts in road projects in whole Germany.[20]

Key Learning of the Section[21]

1. FBV is an "output"-based contract, which specifies the function from the prospective of road users and objective to be achieved, not the process or material requirements.
2. Performance criteria are defined in the contract that is to be fulfilled by the output.
3. The contractor has the freedom to select the process, procedures, technologies, and material requirements.
4. FBV is regarded as a form of PPP in Germany and has been used exclusively in road superstructure.
5. Payment is based on meeting the predefined standard in lump-sum basis. The finance is through the public fund.
6. The functional requirements of FBV are given by the ZTV Funktion-StB (Entwurf).
7. The FBV (parts B and C) is a long-term contract (20–30 years).
8. The risks are allocated based on parts B and C.
9. The selection of the contractor follows the restricted procedures till now. As it is taken as a form of PPP, it could follow the competitive bidding procedures as PPP.
10. The contract is classified into three performance parts: part A (conventional construction contract), part B (function-based construction contract), and part C (function-based maintenance contract).
11. Six projects under FBV have been implemented and two others are in discussion.
12. The Federal Ministry of Transport, Building and Urban Development is striving to implement FBV in whole Germany.

3.5 Public-Private Partnership (PPP)

The PPP project focuses on "output" and the delivery of service rather than the construction methods or procedures; the payment is based on the delivery of a service which meets the performance requirements of the output specification (see pp. 2–28 in [62]). PPP is therefore a project delivery method where the facility constructed should meet the specified standard.

[20] Strassen NRW Press information: www.strassen.nrw.de/service/presse/meldungen/2008/080703-01.html, Accessed on 04.04.2010.

[21] Based on Sect. 3.4; the respective references are presented at the end of this chapter.

Table 3.12 Comparison of elements of PBC with the elements of PPP

Key elements of performance-based contract (PBC) (see this chapter under topic key elements of performance-based contracts (PBC))	Elements of public-private partnership (PPP) (based on the above description of PPP)
PBC is based on the contract instrument which defines the outcome in terms of result as opposed to methods, procedures, systems, or broad categories of work activities	PPP focuses on output and delivery service rather than the construction methods or procedures
The responsibilities are borne by the contractor for the required performance criteria. The contractor is reimbursed depending upon the set of outcomes which require measurable standards (in meters or liters or km) for the development of the project (right pricing arrangement)	The contractor is required to deliver the project; meeting the performance requirements of the output specification and reimbursement to the contractor also depends on these criteria
The contract clearly defines the needs, result, performance criteria, and measures. The acceptable level of performance is identified and defined in the contract. In the case of preferable range of performance, the acceptable range must be mentioned. Incentive criteria are created to motivate the constructor to be innovative to meet or even exceed the level of performance	PPP includes clearly defined results and performance criteria. The completed project should meet the performance criteria defined in the contract in order to be accepted from the owner. PPP also provides the opportunities for the contractor to be innovative, in the sense of using new technology and techniques to reduce costs and deliver quality

Literally from the name, PPP is a cooperation between public and private sector. Different scholars have defined PPP in different ways but the understanding remains the same. According to Van Ham and Koppenjan (2001: 598), "PPP is a cooperation of some sort of durability between public and private actors in which they jointly develop products and services and shear risks, costs and resources which are connected with these products" (see p. 4 in [63]).

The other definition of PPP given by the working group set up by the Northern Ireland Executive to review public-private partnership in 2003 is as follows: "A PPP is generally a medium to long-term relationship between public and private sectors (including the voluntary and community sector), involving the sharing of risks and rewards and the utilization of multi – sectoral skill, expertise and finance to deliver desired policy outcomes that are in the public interest" (Table 3.12; see p. 2 in [64]).

The comparison shows that PPP is a performance-based contract.

According to Lang, Alexander [65] "The performance criteria are described in so called Key Performance Indicator (KPI)." All PPP projects are defied with key performance indicators (KPI) which should be fulfilled by the private sector. The contract defined by the performance indicator seems to be a performance-based contract.

Scott, Sidney and Linda, Konrath (see p. 138 in [6]) state that PPP is based on defined set of agreed-upon performance-based standards which apply the concept of performance-based contract.

From above definitions of PPP, it is clear that PPP is a close cooperation between public and private sectors established to deliver any sort of products or services with an aim of cost reduction with quality. PPP is thus a relation based on shared aspirations and shared risks, rewards, and resources. The project under PPP should be of public interest, for example, infrastructure projects. PPP includes the following key elements (see p. 3 in [66]):

- A medium- to long-term PPP contract between the public sector and private sector
- For the designing, constructing, financing, and the operation of public infrastructure by the private sector (This statement doesn't include the maintenance task. With PPP maintenance is performed by the contractor.)
- With the reimbursement over the life of PPP contract to the private sector for the use of facility constructed, either through the public sector or collected from the users of the facility
- With the facility transferred to the public sector at the end of PPP contracts

Public-private partnership is an alternative for the procurement of the project by public sector using the tax revenues or public borrowing as a source of financing (see p. 5 in [66]). A PPP project involves the financing, design, construction, maintenance, and operation of public infrastructure or a public facility by the private sector under a long-term contract (see pp. 2–28 in [62]). PPP is normally implemented in infrastructure projects, where the public sector plans and designs, whereas the private sector designs in detail, constructs, maintains, and operates the project to earn its investment and profit within the contract period.

Over the last decade, there has been growing trend in delivering public infrastructure through PPP, particularly for services with substantial capital costs.[22] Since the 1990s, there has been a rapid rise of PPPs across the world; the governments of the developing as well as the developed countries are implementing PPPs in infrastructure developments to overcome its financial shortage and improved project delivery; thus PPP is now regarded as the preferred method for public procurement of infrastructure and infrastructure service projects throughout the world (see p. 19 in [68]).

Sectors in which PPPs have been implemented worldwide are listed below (see p. 2 in [69]):

- Power generation and distribution
- Water and sanitation
- Refuse disposal
- Pipelines
- Hospitals
- School buildings and teaching facilities
- Stadiums
- Air traffic control
- Prisons

[22] (Hrab 2003a, b), Shah, Anwar (see p. 139 in [67]).

Table 3.13 Role of different stakeholders in the PPP process (see p. 21 in [69])

Stakeholders	Role
Political decision makers	Establish and prioritize goals and objectives of PPP and communicate these to the public
	Approve decision criteria for selecting preferred PPP option
	Approve recommended PPP option
	Approve regulatory and legal frameworks
Company management and staff	Identify company-specific needs and goals of PPP
	Provide company-specific data
	Assist in marketing and due diligence process
	Implement change
Consumers	Communicate ability and willingness to pay for service
	Express priorities for quality and level of service
	Indentify existing strengths and weaknesses in service
Investors	Provide feedback on attractiveness of various PPP option
	Follow rules and procedures of competitive bidding process
	Perform thorough due diligence resulting in competitive and realistic bidding
Strategic consultants	Provide unbiased evaluation of options for PPP
	Review existing framework and propose reforms
	Act as facilitator for cooperation among stakeholders

- Railways
- Roads
- Billing and other information technology systems
- Housing

3.5.1 Analysis of the Stakeholders in PPP Process

Under PPP, it is understood that public sector and the private partners are the main players; it is also true that there are other players involved in realizing the project. Only two players cannot complete the project delivery. PPP projects are mostly on the infrastructure projects which have direct impact on the general population. The other players involved in PPP projects are listed in Table 3.13 with their responsibilities:

3.5.2 Phases of PPP Project

Every project is completed through the successive process and a procedure; each and every phase is equally important for the success of the project. According to EBST OPP – Vejledning til basiskontrakt, Copenhagen 2005, p. 10 – PPP project

Fig. 3.16 Phases of PPP project

involves five main phases which are time-fixed successive periods; they are presented in Fig. 3.16 (see p. 16 in [22]):

1. *Initiative*
   The initiative phase focuses on defining the needs, specifying the design criteria, and identifying the potential of the project as well as determining whether the project is suitable under PPP (see p. 16 in [22]). It is an important phase as it determines whether or not the particular project is to be implemented under PPP. The wrong decision of this phase may lead to the unsuccessful project delivery.
2. *Design*
   Designing forms the basis for PPP projects; the design phase is a crucial phase because at the design phase much can be done to optimize the value of the finished facility to its end users (see p. 87 in [70]). The design phase includes announcing for bidding and prequalification of the bidder and procurement. The public part chooses the bidder with the most economically lucrative tender and enters into contract, and within this phase the financial aspect is audited (see p. 16 in [22]).
3. *Build*
   The building phase starts with detailed design of the project by the private sector. It selects the most appropriate construction program. The detailed design should be based on the requirements demanded by the owner. Detailed plans for the building, operation, and service are executed and finally the construction works are carried out. The structure of the construction phase under PPP is typically representative of the traditional procurement, which is usually based on the design approach (see p. 69 in [71]). On the completion of the project, the public and private parts evaluate the completed project to ensure the quality as prescribed (see p. 16 in [22]).
4. *Operation*
   It is the phase where the value-of-money argument is tested; payment in PPP is based on the performance in service delivery which is specified in the output specification. On meeting the required level of performance, the private sector is allowed to operate the facility for the agreed contract period to reimburse its investment as well as to earn profit on its investments (see p. 93 in [70]). Maintenance and operation run parallel.
   Maintenance
   In PPP project, the contractor has the responsibility to carry the maintenance of the project for the contracting period. During the operation phase, the service and maintenance are carried out (see p. 18 in [22]). During maintenance, the predefined standard has to be maintained by the contractor.

5. *Transfer*
   The physical facility after the expiry of the contract is handed over to the public sector from the private sector; thus, the facility finally stays in the hand of the public. Sometimes the facility also remains with the private sector depending on the model (see p. 16 in [22]).

3.5.3 Benefits of PPP

PPP procurement brings variety of benefits to the government. PPP makes use of the private finance for the development of public infrastructure; this way the government achieves financial benefits which would boost the economic efficiency and effectiveness of the country. The benefits of PPP are described below (see p. 7 in [72]):

1. *Enhance government's capacity to develop integrated solution*
   With traditional contracts, the large projects are divided into subparts and managed as separate projects because of the lack of sufficient funding from the part of government. This problem is an obstacle for the integrated solution option, but with PPP procurement method, the financing is coming in the sufficient amount from the private sector, through which the government can switch to develop integrated solution for the project development (pp. 7–9 in [72]).
2. *Reduce the cost to implement the project*
   PPP procurement helps to cut the costs compared to the conventional costs or by delivering higher quality with the same costs; this is achieved as with PPP the innovative and creative approach are implemented. The costs for the private sector is lower than public sector in designing and construction works because the aim of private sector is to gain more profit which could be achieved only if the costs could be kept lower. In PPP, cost reduction is achieved through the attribute of synergies, economies of scale, and reduction in life cycle costs (NS, 2000) (see p. 8 in [72]).
3. *Reduce the time to implement the project*
   In conventional procurement method, the projects are broken into smaller units and each unit is carried over an extended period; also acquiring budget for the public projects from government needs longer time. According to NS (2000), cost reduction with PPP is achieved because it (see p. 8 in [72]):

   - Enables design and construction to be undertaken concurrently rather than sequentially
   - Incorporates incentives in the project that rewards the private partner for on-time completion of the project
   - Reduce the number of times a government project or proposal goes out to tender
   - Discourage the temptation to make ongoing changes to the project design, which can cause both delays and create costs

- Time is money; the delay in the implementation results in the additional costs to the private sector which it always wants to avoid.

4. *Transfer certain risk to the private project partner*
The feature of PPP is the managing and transferring of the risks as per the definition of the PPP procurement. With PPP, the strategy is developed to transfer the risk to the party who can best manage the particular type of risks more efficiently (see p. 8 in [72]). The earlier the risks are identified, the more efficiently they can be managed. With conventional procurement methods, most of the risks are borne by the public sector, while with PPP, risks are sheared between public sector and private sector.
5. *Attract larger, potentially more sophisticate bidder to the project*
Depending upon the size and scope of the PPP projects, they are very interesting for the private construction entity and therefore increase the potential bidders and the intensity of the competition. Mostly the infrastructure projects are costly, which need potential constructors. PPP projects when built and operated efficiently and efficiently would result in higher monetary benefits (see p. 9 in [72]).
6. *Access skill, experience, and technology*
PPP procurement is based on creative and innovative approach, where the performance level should be met for the payment. Government can thus gain new skills, technology, and knowledge from the implementation of PPP with such aspects (see p. 9 in [72]). This will help the government sector to gain expertise beyond the aspects associated with conventional procurement methods.
7. *Achieve value for money (VfM)*
Achieving VfM is the main feature of PPP. It is defined as the optimal combination of whole life costs and the quantity to meet user's needs and requirements and is achieved in PPP through (see p. 6 in [73]):

    - Risk transfer which allocates risks optimally between the public and private sectors
    - Long-term nature of contracts (which embodies whole life costing)
    - The use of output specification which allows bidders to innovate
    - Competition that provides fair value of the project
    - Performance-based payment mechanism
    - Private sector management expertise and skills

3.5.4 Features/Characteristics

PPP is widely used in the infrastructure development across the world because of its diverse features and characteristics which bring about several benefits for the private and the public sector. The key features and characteristics are as follows (see p. 5 in [73]):

1. *Relationship between public and private sectors is based on partnership*
PPP is a cooperation between public and private sector as per the definition. The contract is based on mutual benefits. The relationship is based on shared aspiration

between the public and the private sector to deliver the project in the required level of output.

2. *Public sector procures specified outputs or outcomes of a service for a concession period*
   Reimbursement is based on the performance level of output and service level. With PPP requirement of the public sector should be fulfilled by the private sector delivering the project as specified level of results. The required service outcomes are defined in terms of output.
3. *Private sector determines the required inputs to achieve the specified output*
   The public sector has no concern over the material used for the construction rather the performance level of output. Private sector thus has freedom to choose the inputs in terms of material, process, personnel, technology, and procedure. The private sector is given latitude to introduce innovation into their designs and development to reduce overall costs (see p. 5 in [73]). But it has to be careful about the required standard of the output.
4. *Payment for services is based on predetermined standards and performance*
   The payment under PPP as being performance-based contract is made only for the services which meet the contract standards.[23] So, private sector should meet the predefined level to be allowed to start the operation which would ultimately lead to the collection of payments.
5. *Promotes "maintenance culture"*
   During operation phase, the contractor is also responsible for the long-term maintenance of the assets till the handover of the project. Handover takes place after the inspection from the public authority (see p. 5 in [73]). Thus, PPP develops the maintenance culture.
6. *Integration of design, construction, finance, maintenance, and operation*
   All these are the phases of PPP, which is integrated as a package to undertake the project delivery.
7. *Transfer of assets*
   The asset remains in the hand of private sector during the period of contract. But it is transferred to the public by the expiry of the contracts. In some cases, it can also remain in the hand of private sector for some more years depending upon the model (see p. 5 in [73]).
8. *Optimal sharing of risks*
   Sharing of risk is the most important feature of the PPP. Risk is allocated to the party who is best able to manage it. Thus, risks are divided between the public and private sectors. (See Sect. 3.5.7 for detail explanation on risk allocation.)
9. *Whole Life Cycle Costing*
   PPP projects are awarded based on the lowest total costs over the contracting period compared to lowest construction in case of the traditional procurement (see p. 6 in [73]). Thus, the whole life costs are considered, not only the construction costs.

[23] Contract information Bulletin: www.usaid.gov/business/business_opportunities/cib/pdf/cib91_18.pdf, Accessed on: 06.04.2010.

Table 3.14 Role and responsibilities of private sector and public sector in PPP (see p. 10 in [73])

Parties	Responsibilities
Special purpose vehicle (SPV)	Raising the funds to develop and maintain the assets
	Making payments to the subcontractors, financiers, and other creditors
	Delivering the agreed services to the public sector according to the levels, quality, and timeliness of the service provision throughout the contract period
	Ensuring that revertible assets/facilities are transferred in the specified conditions (good working order) to the public sector at the end of the concession period
Financiers	The financing of the project is provided by a combination of equity investors and debt providers
Construction contractors	To carry out construction works according to the contract with the SPV
Facilities management operator	To carry out comprehensive facilities management of the assets according to the contract with the SPV
The public sector	Identifying, assessing and prioritizing projects for implementation via PPP
	Preparing and managing the projects for competitive bidding process
	Providing clear objectives and scoping of the project, output specifications, payment mechanism and KPIs
	Ensuring equitable and optimal allocation of risks
	Contract management and performance monitoring
	Safeguarding public interests

3.5.5 Structure of the PPP Project

The structure of PPP projects presents the clear picture of all the involved private parties with their clearly defined tasks and risks of the projects. The main parties are (see p. 9 in [73]):

- The SPV created specifically for the project
- Financiers
- Construction contractor
- Facilities management operator
- The public sector (procuring authority)

The typical PPP project structure is shown in Fig. 3.19 which presents the activities carried out by each involving party.

SPV is a separate legal entity or a company established normally by the private sector consortium to undertake the activity defined in a contract between SPV and its client, in this case (PPP) public procurer, and helps the private sector in case of bankrupt because SPV is limited in nature (see p. 109 in [75]). In practice, SPVs are often group of banks and other financial institutions that combine and coordinate the use of their capital and financial expertise (Table 3.14, Fig. 3.17; see p. 8 in [76]).

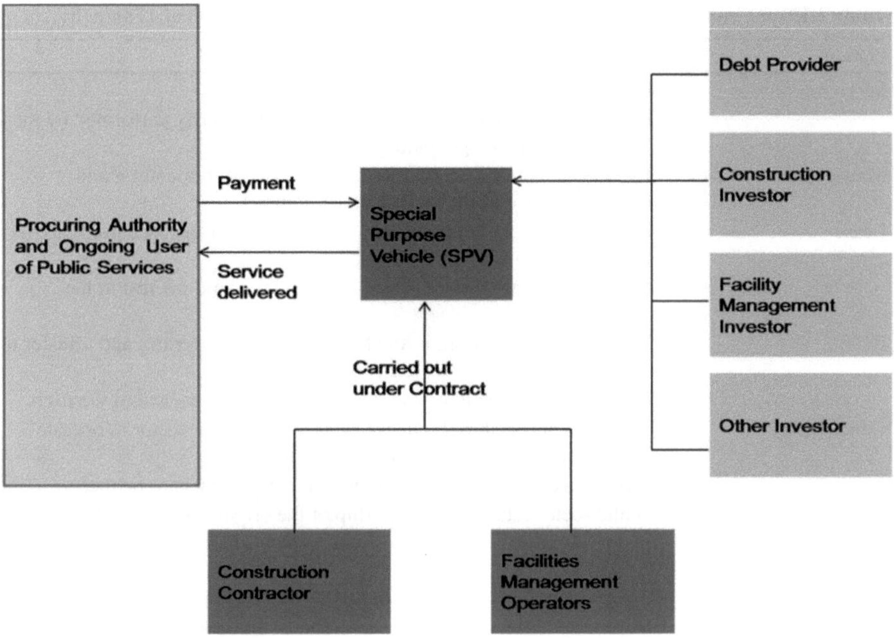

Fig. 3.17 Typical PPP structure (see p. 9 in [73])

3.5.6 *Forms of PPP*

PPP has been used in the variety of forms across the world; it takes forms such as design, construct, and maintain and build, own, operate, and transfer; the choice of the particular form depends on the objectives of the government, nature of the project, availability of the fund, and the expertise that private sector can bring (see p. i in [77]). Each project and each PPP form have their unique characteristics. For the successful project completion, the form of PPP should match with the type of project (Table 3.15).

Contradiction: *Design-build has been considered as a form of PPP. According to the definition of Zimmermann, Josef,* (see pp. 7–28 in [22]) *PPP includes finance, construction, renovation, management or maintenance, and service (operation). Design-build includes design and build aspects. Therefore, design-build cannot be a form of PPP.*

3.5.7 *Risk Allocation*

"Risk is a measure of the probability and consequence of not achieving a defined project goal" (see p. 743 in [79]). It can also be regarded as a notion of uncertainty. Risk directly affects the achievements; so the success of any project depends on the ability

Table 3.15 Forms of PPP and their characteristics (Adapted from, Eggers et al. [78], p. 5)

Option	Characteristics
Service contracts	Just a service contract
	Private entity provides service which was previously performed by the government and get paid for it
Management contract	The private entity is responsible for all aspects of operation and maintenance of the facility under contract
Lease	The private partner operates and maintains the project according to the terms of lease in the contract
	The government grants a private entity a leasehold interest in the project
BOT	The private sector designs and builds the assets, operates and transfer to the public sector after the expiry of the contract
	The contractor is paid by the government or from the fee of the users
Concession	Government transfers exclusive right to the private sector to operate and maintain for a long period of time
	Performance requirements are set by the government to be met
	Public sector retains the ownership of the original assets while the private is the owner of any improvements made during contract period
Divesture	Government transfers assets in part or full to the private sector
	Government includes certain conditions with transfer to ensure that public is served properly
DB	Private sector designs and builds according to the requirements of the government
	Government assumes responsibilities after construction and operates it
DBFO/M	The private sector designs, builds, finances, operates and sometimes also maintains the assets
	The project is transferred to the government after the contract period
DBM	The private sector is responsible for the design and build as well as maintenance for the contract period
	The responsibility of operation is with public sector
DBO/DBOM	The private sector designs and builds the facility
	After the completion of building, private sector operates for the specified time with the option of maintenance
BOO/BOOT	The private entity retains the responsibility of building, financing and operating from the government
	The private entity is not required to transfer the facility back to government
	With option BOOT, the entity is transferred to the government after the contract period

to manage it. The risks associated with PPP projects are the result of the complexity on the arrangement in terms of documentation, financing, taxation, technical details, sub-agreements, etc. involved in a infrastructure projects; the nature of risks however differs over the duration of the project (see p. 109 in [80]). The main objective of risk allocation is to manage risks in the best possible way to achieve the value of money. As discussed earlier, in PPP, the risks are shared among the partners who can best handle

Table 3.16 Types of risks associated with PPP (see p. 111 in [80])

Risks	Risk description
Technical	Because of engineering and design failure
Construction	Because of failure in construction techniques, increase in costs, and delay in construction
Operating	Because of higher operating and maintenance costs
Revenue	Because of traffic shortfall, failure to extract resources, violation of the prices and demand for products and services sold, and revenue deficiency
Financial	Because of inadequate hedging of revenue streams and financial costs
Force majeure	Because of war and other disaster
Regulatory/political	Due to legal change and unsupportive government policies
Environmental	Because of adverse environmental impacts and hazards
Project default	Because of failure of the project resulting from the combination of above defined risks

them with minimal costs. Public and the private sectors are the two main partners associated with PPP; therefore, the risks are shared among themselves.

Risks can never be eliminated but could be allocated; so it is borne in an agreed-upon way (see p. 5 in [81]). The risks could be managed following the following strategies (see p. 6 in [82]):

Risk avoidance – the action is taken to eliminate a particular type of risk.
Risk mitigation – effort to minimizing the effect of risk after its identification.
Risk transfer – risk is transfer to the party who can manage it the best way.
Risk retention – risks are retained if its managing costs more than its effect or no solution are found.

There are several risks associated with the infrastructure investments at the same time with the PPP projects which are listed in Table 3.16.

It is important to identify, analyze, and classify the risks in the early stage as possible to manage them properly to avoid the possible loss. Risk management aspect should be implemented properly. Specially for PPP, where the majority of the tasks are carried out by the private part, the possibilities of using the private know-how to reach a higher assessment and detection potential of the risks are obvious (see pp. 4–2 in [83]). Know-how of the private sector is an important weapon to manage risks.

Risks can be divided in two categories: the risks that can be calculated and the other is the risk that cannot be calculated like political changes, third party intervention, or weather hazards. The private sector should not take the risk which cannot be calculated and controlled; it is also not allowed according to VOB/A (in Germany).[24] If the risks are identified early, it would be easier to handle than the risks identified in the late stage.

[24] VOB/A §9 (2): The contractor is not allowed to be imposed with any abnormal risk of circumstances and incidents on which he has no influence and where he cannot estimate their effect on price and terms (Source: Zimmermann, Josef (see p. 13 in [83]).

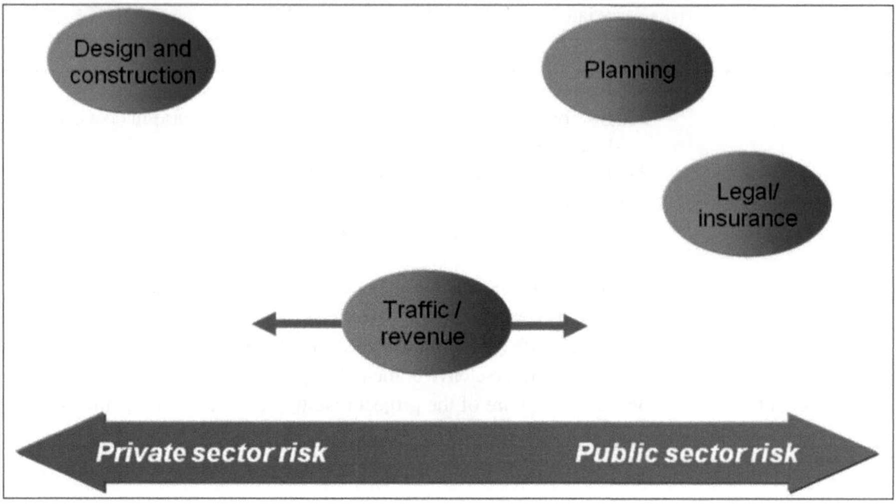

Fig. 3.18 Risk allocation by public and private sector (Based on Seibert, Timlan [84], slide 6)

Distinction can be made between transferable and nontransferable risks.[25] The nontransferable risks such as worst weather conditions or amendment of rules because of the change of government should be handled by the public part. One of the most important features of PPP is the transfer of risk from public to private partner; thus, an appropriate risk transfer strategy needs to be developed as a part of the planning process of the PPP project, in which risk best managed by the private partner is transferred to private and the risk best managed by the public is managed by the public (Fig. 3.18; see p. 8 in [72]).

Risk allocation between the parties however may be different for different PPP projects; the requirement is to transfer substantial risks particularly in the area of design, construction, and operation of PPP projects (see p. 183 in [72]). The risk categories could be further divided into risk profile as small units and then can be transferred to the respective party who can best handle them, to be more specific and effective. In principle, all risks can be transferred to the private part with the provision of economical compensation, but the risks that could be insured are to be transferred to the private part (see p. 13 and 14 in [22]).

3.5.7.1 Public Sector Comparator (PSC)

PSC involves valuing and allocation of risks in order to achieve saving for the whole system. Before undertaking a PPP, a government should ensure that PPP will deliver better value for money (VfM), which requires an ex ante comparison of the VfM of both the PPP and the traditional procurement in every case where the government

[25] OBB: PPP zur Realisierung öffentlicher Baumassnahmen in Bayern, Teil 3, München 2006 (see p. 13 in [78]).

wants to use PPP; therefore, PSC is taken as an instrument that the government can use to conduct the comparison (see p. 69 in [85]). PSC serves as a reference of a hypothetical risk-adjusted costing, by the public sector as a supplier, to an output specification produced as part of a PFI procurement exercise (see p. 104 in [86]). PSC thus ensures that the PPP would bring more benefits than the traditional procurement methods in terms of VfM. It is regarded as a reference for the control of various project activities and would be an important instrument to manage risks.

3.5.8 PPP Procurement

The procurement process includes tendering and contracting which are essential for establishing the overall frameworks for the resulting project and thus for achieving desired efficiency gains and optimal outcomes; PPP procurement must be carefully designed and supported by adequate quality assurance mechanism (see p. 163 in [87]). PPP is a long-term contract with large amount involved, so the procurement procedure is complex and lengthy. The focus of the contract should be on the result of the outcomes rather than how to build because PPP is performance-based contract.

3.5.8.1 The PPP Project Process

The PPP project procurement process begins with the prequalification of the bidder, selecting the most competitive bidder and contracting with the selected bidder, and ends with the financial review. The PPP offering could be limited offering, offering after negotiation, or competitive dialogue (see p. 27 in [88]). The PPP project procurement process is shown below in Table 3.17 (Fig. 3.19).

The main players of PPP are government agency, Shareholders, Building Contractor, Facilities Manager, Special Purpose Vehicle (SPCo/SPV/SPC), and Financiers (financing institutions like banks) are bonded together as shown in Fig. 4.1 with a written contract.

1. *Limited Offering*
   From its name, it has many restriction, rules and regulation in the process of procurement and not opened to all public. This type of procedure is followed in the case when the cost linked to public offering is very high. This type of offering is normally not in practice in Germany and other European countries for the PPP procurement.[26] High transaction costs results when the number of bidders is too high, so in order to reduce such high cost in the process procurement process, limited offering are used. But the companies are not much interested in participating in project with limited offerings; it is normally implemented in the countries where PPP is new to gain experiences.[27]

[26] Based on Kohl, Bernhard (see p. 27 in [89]).

[27] Based on (see p. 27 in [90]).

Table 3.17 PPP project process (see p. 27 in [74])

Phase	Activity
Prequalification	Objectives and priorities; *Define PPP Service*
	Tech., economic, commercial, financial, environmental reviews
	Pre-feasibility study
	Bid evaluation criteria determined
	Prequalification notice issued
Tendering	Preparation of bid documents and contracts
	Preparation for detailed Information Memorandum for lenders
	Issue request for bids to pre-qualifiers
Bid negotiations	Prepare for bid evaluation
	Receipt of bids
	Evaluate bids
	Negotiate with preferred bidder and confirm funding proposals
SPCo formation fin. structure	Detailed negotiations on concession, PPA, FSA, etc.
	SPCo formation and equity allocation
	Approve licences, taxation regime, etc.
	Negotiate Govt. undertakings
	Concessionaire negotiates loan agreements
Financial close	Complete all necessary documentation and conduct precedent

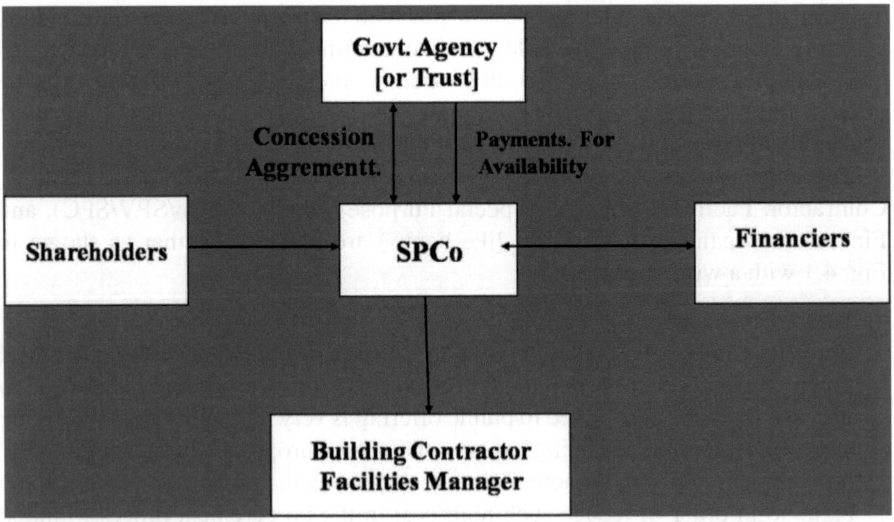

Fig. 3.19 Typical PPP contractual structure (see p. 36 in [74])

2. *Offering After Negotiation*
   Offering after negotiation is normally used in "in exceptional cases, when the nature of works, supplies, or services or the risks attaching thereto do not permit

prior overall pricing."[28] The projects possess various nature and characteristics and it is in some cases very complicated to price the project; in such case the authority addresses certain bidders and discuss on the related terms and conditions. The performance criteria are discussed without price setting in the initial phase of tendering. This initial negotiation helps the authority in setting the price but further negotiation in the later tendering process is always possible. Offering after negotiation can be undertaken with or without previous announcement but it can be undertaken without previous announcement only in the exception cases.[29] This implies that such type of negotiation is applicable when the particular type of project is new and without prior experience because in such case it is very difficult to determine and fix the costs of project. In some case the authority may find the suitable bidder during initial negotiation, in this case the authority may select the particular bidder without further announcement.

3. *Competitive Dialogue*
   It is used in awarding the public project by the public authority if the project is complex and the contracting authorities need to discuss all aspects of the proposed contract with the bidder; the features of competitive dialogue are as follows (see p. 3 in [91]):

   - Selected suppliers/bidders are invited for the dialogue.
   - Results to meet the objective of the project are identified and defined.
   - Economically advantageous tender criteria are followed to award the contract.
   - Dialogue may be followed in successive stage to reduce the number of bidders.
   - Explicit regulations are followed on post-tender discussion.

   Competitive dialogue is used over a limited offering and offering after negotiation in the case where government cannot objectively define technical rules capable of satisfying their needs and objectives and also not able to specify the legal and/or financial makeup of the project (see p. 4 in [91]).

In reality, most PPP or PFI projects are too complex to use open or restricted procedures; so the authorities are using negotiated contract like competitive dialogue because of its diverse features (see p. 1 in [92]). It has been adapted throughout the EU and been in application in countries like France, Germany, Denmark, and Malta (see p. 1 in [92]; Fig. 3.20).

The authority needs to publish a contract notice setting their objectives and goals. The selection procedures assume after the expression of interest from the various construction contractors. On the completion of selection of the competitive candidates, the invitation to dialogue is carried out. The solutions for the projects, the objectives, and

---

[28] Based on Construction and Plant directive 2004718/EC, Article 30(1) (b): Zimmermann, Josef: Script for lecture "Project Delivery Systems", Lehrstuhl für Bauprozessmanagement der TU München, Ausgabe 01/2009, p. 27.

[29] Based on Konkurrencestyrelsen: Bilag 9 – reglerne for anvendelse af udbud efter forhandling, www.ks.dk, Accessed on 09.12.2007: Zimmermann, Josef (see p. 28 in [83]).

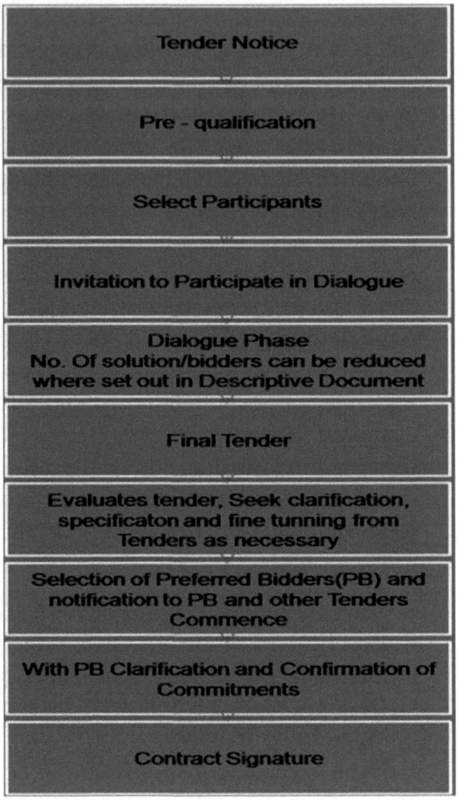

Fig. 3.20 Competitive dialogue procedure (see p. 8 in [91])

needs are discussed during the dialogue, after which the candidate are reduced to a certain number to ensure better selection procedure and effective competition. After dialogue, final tendering phase starts. As soon as the authority exactly knows what he wants, the candidates selected from the dialogue are asked to submit the final offer. The final tender should contain all the elements required to achieve the performance and output required. Sometimes post-tender discussion can also be carried out. The tender is then evaluated according to the defined award criteria of the authority, and the most economically advantageous tender is then selected (see p. 5 in [91]). The payment starts after the project is set into operation either from the public sector or collected from the user's fee (see p. 30 in [88]).

Key Learning of the Section[30]

1. PPP is a performance-based contract and focuses on "output" rather than the methods or procedures.
2. PPP is a cooperation between public and private sector.

[30] Based on Sect. 3.5, the respective references are presented at the end of this chapter.

3. PPP is widely used in variety of infrastructure including road, hospital, railways, and sanitation.
4. The contractor performs detail design, construction, and operation/maintenance tasks and transfers the project after the expiry of the contract.
5. PPP is a long-term contract (20–30 years).
6. Private sector bears the risk of design and construction.
7. Selection of contractor under PPP is normally though competitive bidding.
8. A part or the whole financing of the project is shouldered by the private partner. Normally, PPP has been developed to use the private finance in the infrastructure development.
9. Payment is based on the delivery of a service which meets the performance requirements of the output.
10. The payment to the contractor is collected directly from the users or paid by the public sector.
11. Since the 1990s, there has been rapid rise of PPP all over the world.

References

1. Molenaar KR, Yakowenko G. Alternative project delivery, procurement, and contracting methods for highways. Reston: ASCE; 2007.
2. National Research Council (US), Committee for Oversight and Assessment of U.S Department. Process in improving project management at the department of energy. 2003 assessment. Washington, DC: National Academy Press; 2004
3. White B, Newcomer KE. Getting result: a guide for Federal leaders and managers. Vienna: Management Concepts; 2005.
4. Testa MF, Poertner J. Fostering accountability using evidence to guide and improve child welfare policy. New York: Oxford University Press; 2010.
5. Federal Facility Council, Standing Committee on Organization and Administration. Federal facilities beyond the 1990s: ensuring quality in an era of limited resources – summary of symposium. Washington, DC: National Academy Press; 1997.
6. Sidney S III, Konrath L. Performance based contracting: a viable contract option?, Summary of conference, U.S. and international approaches to performance measurement for transportation systems, conference proceedings 44, Washington, DC.
7. Cibinic, John. Performance-based contracting, guide. Washington, DC: US Department of Energy; 2000.
8. NCHRP Synthesis 389: performance-based contracting for maintenance. Washington, DC: Transportation Research Board Publication; 2009.
9. Parmenter D. Key performance indicators, developing, implementing, and using winning KPIs. Hoboken: Wiley; 2007.
10. Performance indicators for road sectors: summary of the field test. Paris: OECD; 2001.
11. Zietlow G. Cutting costs and improving quality through performance-based road management and maintenance contracts –Latin America and OECD experiences, Birmingham, April 2004.
12. Blokdijk G. Agreement 100 success secrets, SLA, service level agreements, service level management and much more. Australia: Emereo Pty Ltd; 2008.
13. Bangemann TO. Shared services in finance and accounting. Aldershot: Gower Publisher Ltd; 2005.
14. Atkin B, Brooke A. Total facilities management. Oxford: Blackwell; 2000.
15. Camarinha-Matos L, Paraskakis I. Leveraging knowledge for innovation in collaborative networks, Greece, October 2009.

16. Click RL, Duening TN. Business process outsourcing, the competitive advantage. Hoboken/Canada: Wiley; 2005.
17. Brown PC. Implementing SOA – total architecture in practice. Boston: Addison-Wesley; 2008.
18. Glavinich TE. Contractor's guide to green building construction. Hoboken: Wiley; 2008.
19. NCHRP Report 428: Guidebook to highway contracting for innovation: the role of procurement and contracting approaches in facilitating the implementation of research findings, Washington, DC, 1999.
20. Cushman, RF, Cushman KM, Cook SB. Construction litigation: presenting the owner. USA: Asphen Publishers; 1990.
21. Stankevich N, Qureshi N, Queiroz C. Performance-based contracting for preservation and improvement of road assets. Washington, DC: The World Bank; September 2005 (updated August 2009).
22. Zimmermann J. Project delivery systems, lecture note in Lehrstuhl für Bauprozessmanagement und Immobilienentwicklung an der Technischen Universität Müncen, issue January 2009.
23. Performance based contracts: global transport knowledge partnership. www.gtkp.com/theme.php?themepgid=118. Accessed 21 Mar 2010.
24. Stankevich N, Qureshi N, Queiroz C. Performance-based contracting for preservation and improvement of road assets. Washington, DC: The World Bank; September 2005 (updated August 2009).
25. Burningham S, Stankevich N. Why road maintenance is important and how to get it done, The World Bank Document, TRN-4, June 2005.
26. The Glossary of Federal Technology. www.nextgov.com/the_basics/tb_20080605_4625.php. Accessed 31 May 2010.
27. Zietlow G. Implementing performance-based road management and maintenance contracts in developing countries – an instrument of German technical cooperation, Eschborn, November 2004.
28. Pakkala P. Performance-based contracts – international experiences, Finnish Road Administration, Presentation at the TRB workshop on performance-based contracting, Washington, DC, April 2005.
29. ADB: Performance – based maintenance by contracts, Supplementary Appendix A, P.1. www.adb.org/Documents/Supplementary-Appendixes/39676/Supplementary-Appendix-A.pdf. Accessed 24 Apr 2010.
30. Uthus L. The perfect performance based contract – how should it be?. www.nvfnorden.org/lisalib/getfile.aspx?itemid=607. Accessed 31 May 2010.
31. Shrestha HR. Sustainable approach to road transport maintenance practice for effective service delivery. Policy paper 24, Kathmandu. September 2006.
32. Procurement of performance-based management and maintenance of roads (Output-based service contract). The World Bank report, Revised March 2003.
33. Brian C, Fekpe E, Gopalakrishna D. An evaluation of the use of innovative performance specification in highway construction: final report, Washington, DC, February 2003.
34. Forsthoffer W. Reliability optimization through component condition monitoring and root cause analysis. Oxford: Elsevier; 2005.
35. Zietsman J. Performance measure for performance based maintenance contracts, research of Texas Department of Transportation. www-esd.worldbank.org/pbc_resource_guide/Docs-latest edition/cases-and-pdfs/ZietsmanTexas.pdf.pdf. Accessed 10 Apr 2010.
36. Procurement of performance-based management and maintenance of roads. Washington, DC: The World Bank. February 2002.
37. Schmerbeck R. Pilotprojekte mit Funktionsbauvertragen – Vertragliche Sicherung der Oberflächeneigenschaften, München. http://www.vsvi-hessen.de/download/20051102/vsvi2005_11_02schmerbeck.pdf. Accessed 26 Mar 2010.
38. Pauli C. Schriftenreihe Bauwirtschaft – Tagungen und Berichte, IBW-Symposium, Innovative Abwicklungsformen für Bauprojekte: Partnering with PPP. Kassel, 2006.
39. Alfen HW, Leupold A. Offering new and integrating opportunities for the private sector, Public Private Partnerships (PPP) in the German Public Real Estate Sector. www.europe-re.com/system/main.php?pageid = 2242&articleid = 9378. Accessed 26 Mar 2010.

40. Beckers T. Die Realisierung von Projekten nach dem PPP-Ansatz bei Bundesfernstraßen, Dissertion, TU Berlin, Fakultät Wurtschaft & Management, Berlin, 2005.
41. Nösler I. Oberbaubemessung im Rahmen neuer Bauvertragsformen (Funktionsbauvertrag), Brüssel. www.nynas.com/upload/docs/Bitumen/NOAH.pdf. Accessed 26 Mar 2010.
42. Güngerisch A. Neue Bau- und Beschaffungsverträge, USIC seminar, Bern, 15 October 2008.
43. Verdi: PPP in Strassenbereich, April 2007.
44. Racky P. Der Funktionsbauvertrag aus betriebswirtschaftlicher Sicht, VSVI – Seminar "Funktionsbauverträge", Friedelberg, April 2010.
45. Eifert, H. Funktionsbauverträge – ein Modell der Zukunft? Köln, April 2004.
46. Racky P. PPP in Hessen – auch im Kommunalen Strassenbau?, Summary of the presentation, Hessen, March 2003.
47. Alfen Consult GmbH Weimer. PPP- Businessmodelle im Überblick, Weimar, March 2008.
48. Altmüller P. Schriftenreihe Bauwirtschaft – Tagungen und Berichte 6, Kassel, April 2009.
49. Schmerbeck, Löcherer. Vermerk Bundesautobahn A 93 Süd Rosenheim – Kiefersfelden AS Brannenburg – AS Kiefersfelden, Vermerk über die Erfahrungen bei der Vergabe und Bau, Autobahndirektion Südbayern, München, June 2008.
50. Nösler I. Pavement design in Rahmen neuer Bauvertragsformen, Brüssel, tis 10/2003.
51. Gerdes E. PPP-Ansätze für Fernstrasse, Dissertation an Universität Hamburg, Hamburg, Germany, December 2007.
52. Presentation of the Münchner Massivbau-Kolloquium Sommersemester 2009, München.
53. Wiederspahn M. Brückenbau: 10. Symposium Brückenbau in Leipzig, February 2010.
54. Beckers T, Hirschhausen CV, Klatt JP. Reformbedarf bei den Bundesstrassen und das potential des PPP-Ansatzes, Berlin, May 2009.
55. Ressel W. Universitaet Stuttgart, Kolloquium- Technische Absicherung des Regelwerks, Privatfinanzierter Strassenbau – Chancen und Risiken, Stuttgart, 2005.
56. Beckers T, Klatt JP. Zeitliche Homogenisierung und Berücksichtigung von Risiko im Rahmen von Wirtschaftlichkeitsuntersuchungen, Berlin, November 2009.
57. Heer N. Diplomarbeit Summary. www1.tu-darmstadt.de/fb/bi/ifv/sw/stud/kfd/n_heer.pdf. Accessed 2 Apr 2010.
58. Schaefer S. Der Funktionsbauvertrag, Presentation11, Lehrstuhl Bauproyessmanagement und Immobilenentwicklung Technische Universitaet Muenchen, Munich, Germany, Ausgabe 02/2010.
59. Forschungsgesellschaft für Strassen – und Verkehrswesen, Arbeitsgruppe Fahrzeug und Fahrbahn: ZTV Funktion-StB 01, Koeln, 2001.
60. Dreher A. Pilotprojkte mit Funktionsbauvertraege – Erwartung der Strassenbauverwaltung, Koblenz, Mai 2003 (Strasse + Autobhan, 54 Jahrsgang).
61. Beckers T., Hirschhausen, CV., Klatt JP. Reformbedarf bei den Bundesstrassen und das potential des PPP-Ansatzes, Berlin, June 2006.
62. Ballati D. Privatizing governmental functions. New York: Law Journal Press; 2004.
63. Hodge G, Greve C. Public – private partnership, learning from international experience. Cheltenham: Edward Elgar; 2005.
64. Geddes M. Making public private partnerships works, building partnerships and understanding cultures. Aldershot/Burlington: Gower Publishing; 2005.
65. Alexander. Investor – Hochschule – Bauindustrie, Kolloquium 2006, Muenchen, under the topic PPP.
66. Yescombe ER. Public private partnerships, principles of policy and finance. Oxford: Butterworth-Heinemann; 2007.
67. Shah A. Public service delivery. Washington, DC: The World Bank Publication; 2005.
68. Asian Development Bank. Facilitating Public-Private Partnership for accelerated infrastructure development in India, Workshop report, India, December 2006.
69. Asian Development Bank. Public-Private Partnership handbook. Manila: Phillippines, 2008.
70. Robinson H, Carrilo P, Anumba C, Patel M. Governance and knowledge-management for public-private partnership. USA: John Wiley and Sons Publication; 2010.

71. Construction cost management; learning from case studies. Oxon/USA/Canada: Taylor & Francis; 2008.
72. Akintoye A, Beck M, Hardcastle C. Public-Private Partnership: managing risks and opportunities. New York: Wiley Publication; 2003.
73. Public Private Partnership (PPP) guideline: Public-Private Partnership Unit, Prime Minister Department, Putrajaya, Malaysia, 2009.
74. Blaiklock TM. Public – Private Partnership ("PPP") – the advantages and limitations, Berne, January 2005.
75. Grimsey D, Lewis MK. Public Private Partnership, the worldwide revolution in infrastructure provision and project finance. Northampton: Edward Elgar; 2004.
76. IMF. Public Private Partnerships, government guarantees, and financial risks. Washington, DC: IMF; 2006.
77. Public Private Partnerships: an introduction, research paper no. 1 2002–2003, Department of the Parliamentary Library, Commonwealth of Australia, 2002.
78. Eggers WD, Startup T. Closing the infrastructure gap: the role of Public-Private Partnerships. New York: Deloitte Research; 2006.
79. Kerzner H. Project management – a system approach to planning, scheduling and controlling. USA: John Wiley and Sons Publication; 2009.
80. Grimsey D, Lewis MK. Evaluating the risks of Public Private Partnerships for infrastructure projects. Int J Project Manag. 2002; 20:107–18..
81. Jordan A. Presentation of managing risks in PPP projects through legal documentation, Jordan, September 2007.
82. Ewers K. Assessing risk for health care IT organizations – a practical overview approach, February 2007.
83. Zimmermann J. Script for lecture "Schlüsselfertiges Bauen", Lehrstuhl für Bauprozessmanagement der TU München, Ausgabe 10/2007
84. Seibert, Timlan. Presentation PPP – The European investment bank experience, Kyiv, March 2006.
85. OECD. Public-Private Partnerships: in pursuit of risks sharing and value of money. Paris: OECD; 2008.
86. HM Treasury. The green book. London: HM Treasury Publication; 2003.
87. OECD. Transport infrastructure investment: options for efficiency. Paris: OECD Publication; 2008.
88. Zimmermann J. Script for lecture "Project Delivery Systems", Lehrstuhl für Bauprozessmanagement der TU München, Ausgabe 01/2009.
89. Kohl B. Interview, München 19.12.2007: Zimmermann J. Script for lecture "Project Delivery Systems", Lehrstuhl für Bauprozessmanagement der TU München, Ausgabe 01/2009.
90. Reichle, M. Interview, München 19.12.2007: Zimmermann J. Script for lecture "Project Delivery Systems", Lehrstuhl für Bauprozessmanagement der TU München, Ausgabe 01/2009.
91. Competitive dialogue procedures: Office of Government Commerce: guidance, London, January 2006.
92. Competitive dialogue: the EU's new Procurement Procedures, Freshfields Bruckhaus Deringer, November 2005.

Chapter 4
Comparative Study on Construction Contracts

Various project delivery methods are in use for the infrastructure development. The aim of each project delivery method is to deliver the physical entity with high quality in possible minimum costs. Cost reduction with quality is the objective of the construction procurements. Every construction method has its own features and characteristics (described in respective chapters), and each follows different procedures (e.g., selection). The types, procedures, and other aspects of different delivery methods are presented in detail in Chaps. 2 and 3. Broadly, the construction contracts are classified as traditional and performance-based. The traditional project delivery methods are most commonly practiced for small projects (e.g., D-B-B in the USA) (see p. 551 in [1]). Design-bid-build is considered as conventional construction in the USA (see pp. 6–26 in [2]). Traditional construction method is not applicable for every situation, but it is still used by the majority of the owners in the construction projects (see p. 16 in [3]). In contrast, Toh, Mun Heng, and TanKong Yam (see p. 152 in [4]) argue that the traditional methods of contracting are being slowly replaced by other alternative form of contracts. The alternative methods could be performance-based contracts like PPP.

Performance-based contracts (PBCs) have been in implementation since about more than 20 years and have been regarded as an effective way of project delivery systems in an effective way in terms of quality and cost-effectiveness (see p. 5 in [5]). PBC has been regarded as the developed and modified form of the traditional contracts improving the drawbacks of the traditional construction contracts. PBCs are constantly modifying and evolving because more experiences are gained and lessons are learned with time (see p. 148 in [6]). Section 4.1 presents the comparative study on the traditional construction contracts and PBCs.

A. Gajurel, *Performance-Based Contracts for Road Projects: Comparative Analysis of Different Types*, DOI 10.1007/978-81-322-1302-4_4,

4.1 Comparison of Traditional Construction Contracts and Performance-Based Construction Contracts

Although both types of contracts have the same objectives to deliver the project with quality and lowest possible price, they differ in terms of procedures and techniques to achieve the defined objectives. According to Molenaar, Keith R., and Yokowenko, Gerald (see p. 21 in [7]), a PBC is different from traditional contract, and the main difference is that the PBC sets the performance expected from the end output of a project, instead of indicating the methods to achieve the end output, whereas in traditional contracts the methods are directed. This chapter aims to present the comparative studies (similarities and differences) between them in detail.

Tradition construction contracts	Performance-based construction contracts (PBCs)
Definition	
Under traditional construction contracts the owner contracts with engineer/architect for the detail design and contract documentation for the works and then enters into a contract with general contractor to construct the whole project [8]. It is defined based on phases and the definition closely points D-B-B. D-B-B is the traditional project delivery system in the USA (see pp. 6–26 in [9])	A performance-based contract is a method of contracting where the owner defines the results it requires rather than the method, materials requirement, and techniques by which the specified results are achieved (see p. 1 in [10]). The contract is defined by performance criteria
Principle	
The contracting agency as an owner normally specifies techniques, procedures, material, as well as the quality of material and the period during which the job should be carried (see p. 1 in [10]) Contract based on input [11]	The owner specifies performance indicators (e.g., no potholes on the contracted section of road) which the contractor is required to meet in delivering the project or services (see p. 1 in [10]) Contract based on output
Payment	
The payment to the contractor is based on the inputs, for example, the amount of concrete or bitumen or working hours, and the payment is based on measurement of quantity of output (see p. 1 in [10]). The payment is based on the unit price on the completion of the project. For the public projects, payment is through the public budget	Payment is done to the contractor in case of his successful meeting or exceeding certain clear predefined performance criteria and based on measured output (see p. 1 in [10]). Payment is based on a fixed price lump sum through uniform installments on meeting preferred performance (see p. 3 in [10]). With PBC the payment is normally through public fund but with PBC contract like PPP, the payment to the private sector contractor could be through public fund or collected directly from the users
Freedom of the constructors	
The owner normally defines the process, procedures, and materials to be used. There is little room for the implementation of innovative approach	The owner doesn't define the process, procedures, and material requirements. The contractor can select technologies, techniques, and procedures to be cost effective (see p. 2 in [10])

(continued)

(continued)

Tradition construction contracts	Performance-based construction contracts (PBCs)
Selection process	
Selection process depends on the lowest price bid approach (see p. 390 in [12]). The low bid approach does not select the contractor based on qualification or proven ability of successful completion of similar projects in the past rather on meeting bid qualification and lowest bid value (see p. 2 in [13])	Selection is based on the best value approach which does not necessarily depend on the lowest bid approach which ensures a high quality output at lower overall costs (see p. 3 in [10]). The "best value approach" means selection process depends on the management capacity, financial capacity, his understanding on the new approach, ability to handle different risks, and past performances on similar projects (see p. 3 in [10])
Risks	
In principle the owner bears the design risks and the risks that are resulted from the deficiencies of the design, except that if designers are negligent, some of their consequent losses may be recovered and the contractor bears most of the construction risks except some are shared with subcontractors and suppliers (e.g., shortage on supply of materials) (see p. 6 and 7 in [14])	Risks under PBC are borne by the party who can best manage them; the contractor bears the risks and responsibilities which it can manage better than the owner (see p. 2 and 3 in [10]). The contractor bears the risks of designing (e.g., the design not matching the required project), construction risks, maintenance risk, and the risk for costs overruns (see p. 9 in [14]). The other unpredictable risks that are beyond the control of contractor like third party intervention and political risk are subject of discussion between the contractor and the owner and should be allocated to that party that can manage properly (see p. 4 in [10]). Risk allocation of different performance-based contracts differs from each other but principle risks such as design risk, construction risk, maintenance risk, and quality risks are borne by the contractor
Duration	
The duration of traditional construction contract is lower than PBC because the contractor under PBC carries more risks and responsibilities (design and construction) and is responsible for the maintenance works every few years (see p. 3 in [10]). The consultant's role usually ends with the completion of the construction phase under traditional method (see p. 9 in [15])	The duration of PBC is normally longer than traditional contracts as the contractors bear more risks and responsibilities and are responsible for the maintenance for some years; the PBC is normally from 3 to 10 years and could be up to 30 years (see p. 3 in [10])
Duties	
The owner contracts separately with designer and constructor. In principle owner sign contract with designer and constructor (see p. 8 in [16])	Work selection, design (work schedule, material selection, selecting staffs, etc.) and construction responsibilities are carried by the contractor (see p. 1 in [17]). Maintenance is also carried by the contractor and in some case operation too (e.g., PPP)

(continued)

(continued)

Tradition construction contracts	Performance-based construction contracts (PBCs)
Output	
In traditional methods material or methods of construction are specified. The constructor is obliged by the methods and materials to deliver the project (see p. 5 in [18])	PBC describes the demands on the quality by standards, the level of output is defined which should be met by the contractor, and the measurable and comparable performance indicators are predefined which would be compared to the achieved performance level. The constructor is obliged to bring at least minimum preferred level of performance for the project acceptance (see p. 5 in [18])
Modification	
Traditional method does not provide opportunities to modify the design late in the design process and in the construction phase (see p. 5 in [13]). The designer and the constructor are different; therefore, the modification in design in late phase is difficult	None of the literature clearly stated that modification of design is possible in late phase of the designing or during construction under PBC. Stankevich, Navaid Qureshi, and Cesar Queiroz (see p. 3 in [10]) argues that the contractor under PBC is not bound regarding "what to do" or "when to do" and "how to do" but he is bound to deliver the specified level of services or outcome and the constructor and designer are the single entity. This means he would have some degree of freedom to modify the design when necessary to deliver the specified level of performances
Performance specification (KPI)	
Traditional contract is based on the specification of the owner regarding time, costs, and quality, and the experts help owner to specify the requirements. The contractor is required to achieve the requirement in terms of time, costs, and quality. Successful achievement of these is fulfilling the contract requirement and expected to be paid regardless the satisfaction of the owner with the results [19]	PBC is based on specifying output, controlling costs, and paying suppliers and maintaining service level. The performance-based contract defines the performance level using SLA (see Sect. 3.1 for detail), as well with PBC; each performance required is defined with key performance indicators (KPI) which the contractor is obliged to deliver in order to be paid and each KPIs are measurable and comparable and have been developed to monitor and measure performance level and to promote continuous improvements [19]
Role of private sector	
The role of private sector is normally very limited and typically ends with the completion of construction (see p. 3 in [20]). The contractor is not involved in the preliminary planning and design phase (see p. 9 in [21])	The contractor undertakes to plan in detail, program, construct, and implement maintenance work in order to achieve specified standard for a fixed price, subject to specified risk allocations; thus, PBC involves more private sector (see p. 21 in [7])

(continued)

(continued)

Tradition construction contracts	Performance-based construction contracts (PBCs)
Objectives	
The objectives of traditional method are:	The objectives of PBC are (see p. 2 in [10]):
Achieving lowest cost by choosing the lowest price bidder (the selection is based on " low bid price")	Cost saving in construction, managing, and maintaining
Separating the responsibilities for design and construction	Managing project with few staffs
Conformance of the owner expectations (based on definition and selection process as described above)	Fewer contacts which reduce the agencies, overhead costs, and staffs
	Delivering better user satisfaction with ensuring the level of performance
	Reacting quickly against the nonperformer and rewarding the better performer
	Applying innovative approach to be cost effective and achieving quality
	Maintaining the project on long-term basis
	Flexibility to react on the changing conditions
Participants	
The owner	The owner
The architect	The contractor
The contractor	
Role of each player	
Role of each player with traditional contracts (see p. 6 and 7 in [22])	Role of each player with PBC (see pp. 1–3 in [10])
Owner:	Owner:
Responsible throughout delivery system	Specifies the output to be delivered by the contractor
Defines the need for the project	Specifies performance indicator to be met by the contractor
Selects the designer and contractor	Pays the costs on completion of the project on meeting the defined standard
Defines and communicates the expected roles of all participants	Selects the contractor based on value approach
Inspection during construction	
Liable to pay the costs	
Designer:	Constructor:
Generates the project drawings and specifications	Plans and designs the projects according to the requirements
Reviews the plans and specification at various stage	Designs in detail and construct the projects
Coordinates the design approval through agency	Maintains the project till the expiry of the contract

(continued)

(continued)

Tradition construction contracts	Performance-based construction contracts (PBCs)
Helps in estimating the probable costs and completes final design	Delivers the project as specified specifications Maintains for a contracting period as predefined standard Handover of the project after the expiry of contract at defined level of standard
Constructor: Selected through the competitive bidding Signs contracts with the owner Constructs the project according to the intent of plans and specifications Submits the progress payment requests and required notice for inspection and attends progress meeting Handover of the project after completion	
Finance	
For the infrastructure projects finance is from government through public fund; the financing is based on the amount of inputs and carried by the responsible authority (see p. 1 in [10])	For the infrastructure projects finance is through the public fund based on contractor successfully meeting or exceeding the performance level predefined and checked against the KPI (see p. 1 in [10]). But with PPP, funding is blend of public and private funds or in some case only by private (see p. 138 in [23]). The private sector is paid back by the government or collected from the users over the contracting period
Payment condition	
The constructor has to fulfill the requirements in terms of time, costs, and quality and is liable to be paid once the construction is completed [19]. Payment is based on unit price	The payment condition is solely dependent on meeting KPI defined in the contract; the contractor is paid periodically on meeting KPI, penalties for noncompliance, and reward for delivery of the project in the defined or exceeding level of performance (see p. 4 in [10])
Error with design	
A construction contractor is paid an additional compensation if errors arise due to errors, omissions, or ambiguities in the plans and specifications (see p. 2 in [24])	The designer and constructor are the single entity with PBC, so such provision is not the subject with performance-based contracts
Cost reduction	
Performance-based contracts in road projects have resulted in cost reduction compared to the traditional contracts, which are unit price contracts. In Latin America, for instance, the cost reduction of about 10 % is achieved from the implementation of performance-based contracts for road projects compared to the traditional construction contracts (see p. 4 and 5 in [25])	

Zietlow, Gunter [26] argues that performance-based contracts have started to replace the traditional method-based contracting method in the field of road maintenance and management more than a decade ago.

4.2 Comparative Study on Different Performance-Based Construction Contracts

The book aims at presenting the analytical study as well as comparative study on the performance-based contracts for infrastructure developments. The analytical study has been presented in Chap. 3. The performance-based contracts that are most widely used according to the literature review and real-world practice are, namely, performance-based road management and maintenance contract (PMMR), public-private partnerships (PPP), and Funktionsbauvertrag (FBV). From the analytical study, it has been seen that these three types of contracts have many similarities as well as differences. The objective of all these contracting methods is to deliver the infrastructure entity with high quality at lowest possible price; however, they differ from each other regarding some processes and procedures.

Principally, they are output-oriented contracts. The functional requirements are predefined to be fulfilled by the contractor. The selection of methods, techniques, and material requirements is dependent on the contractor choice. This chapter aims on presenting the detail on the similarities and differences between them to find out if they are similar to each other in other aspects of implementation. The comparative study is dependent on the analysis of Chap. 3.

Funktionsbauvertrag cannot be compared in general because it consists of contract parts A, B, and C. In short, part A is like the conventional construction contract. Parts B and C are defined with the functional requirements and are long term in nature (for detail, see Sect. 3.4.2).

4.2.1 Comparison of the Concepts

Figure 4.1 presents the concept of PPP, Funktionsbauvertrag, and PMMR. From the figure it can be concluded that PPP performs the tasks of financing and operation, design, construction, and maintenance while Funktionsbauvertrag and PMMR don't include financing and operation aspects. Funktionsbauvertrag and PMMR seem to be similar in concepts as presented by the above figure. The main difference is that Funktionsbauvertrag includes new construction, renovation, and maintenance while PMMR doesn't include the new construction task but includes only renovation, maintenance, and management. With Funktionsbauvertrag, road management is carried by the owner while it is carried by the contractor with PMMR.

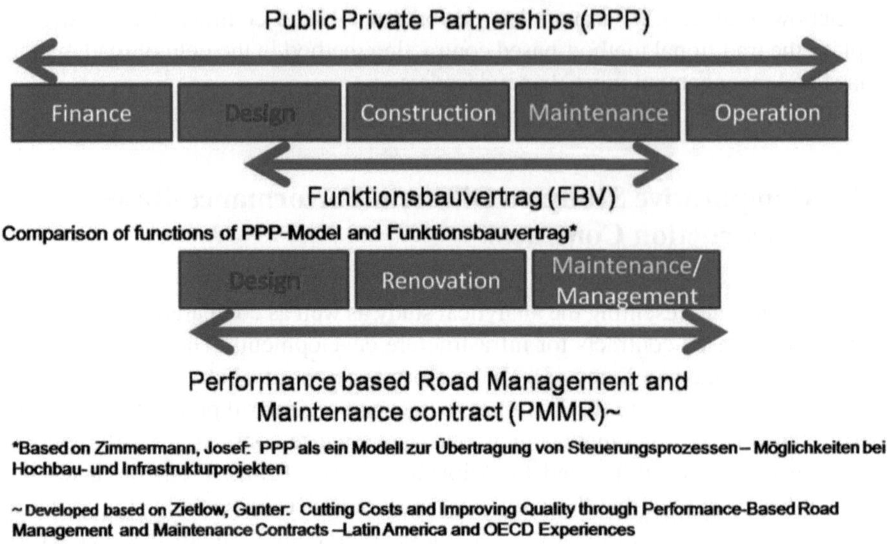

Fig. 4.1 Comparison of functions of PPP model, Funktionsbauvertrag, and PMMR

1. Contract parts

PMMR	The literatures from the World Bank regard PMMR as a single contract. Contract parts within PMMR are not defined
PPP	The PPP contract is regarded as a single contract. However, there could be various contracts with subcontractors, banks, etc.
FBV	The feature of FBV is its contract parts. FBV is divided into contract parts A, B, and C with distinct features and characteristics (see Sect. 3.4.2). It might have simplified the task which could result in the efficiency. It may have resulted in the positive experience with FBV as stated by BMVBS
Conclusion: Dividing into parts simplifies the task which may be easily implemented. These aspects of FBV could be learned by PPP and PMMR	

2. Responsibilities of contractor

PMMR	PMMR expands the role of private sector from simple execution of works to the selection of works, procedures, design, and project delivery (see p. 20 in [27])
PPP	The typical PPP project involves the private party from detail design, selection of work, procedures, construction, maintenance, and operation of the project till the contracting period (depending on the phases of PPP Sect. 3.5.2)
FBV	The responsibility of designing, planning, constructing, and maintenance is transferred to the contractor (see p. 37 in [28])
Conclusion: The available literature of FBV and the World Bank literature regarding PMMR state that design is the responsibility of the contractor. As PMMR and FBV are implemented only for road project, the design task cannot be fully carried by the contractor. The corridor is fixed by the public sector and regarding road resign (road layout, layers thickness, drainage aspects, slopes, road evenness, etc.) is already defined (through VOB in Germany). So the contractor has little role in designing. In regard of PPP, with projects like prison and building, the private sector could be involved in planning and design	

3. Role of the owner (based on the analysis of Chap. 3)

PMMR	Specifies the performance indicators the constructor needs to deliver. Meeting the specified performance level, the owner pays on installment basis and performs the timely inspection. The selection of the contractor through bidding is performed by the owner. Operation and financing of the project are also performed by the owner
PPP	Planning and design are the initial role of the owner. Timely inspection, selection of contractor, and the requirement payment are to be carried by the owner. The owner sets the standard for the project
FBV	Specifies the performance indicators and criteria the constructor needs to deliver. Selection of the contractor and payment to the constructor on the completion of the project maintaining the predefined is carried by the owner with the timely inspection to ensure the standard. Operation and financing of the project are also performed by the owner
Conclusion: All three types of contracts involve the owner in specifying the project, selecting the contractor, making the initial payment, specifying the performance indicators, and paying the costs as agreed. Project management is performed by the owner with FBV but is performed by the contractor with PPP and PMMR. Operation is the task of owner with FBV and PMMR but is the task of contractor with PPP (Table 4.1)	

4. Duration

PMMR	PMMR contract is usually from 3 to 10 years but could go up to 30 years as the maintenance work should be periodically done (see p. 1 in [10])
PPP	The contract period under PPP is normally 25–30 years (see p. 33 in [29])
FBV	The contract period under FBV ranges from 20 to 30 years (parts B and C) (see p. 37 in [28]). Contract part is short-term contract because it is a conventional construction contract
Conclusion: All the contracts are long-term contracts and go till 30 years	

5. Performance indicators

PMMR	PMMR contract clearly defines the performance standards to be delivered by the contractor. The objective of the performance indicator is to satisfy the set goals and comparison and control during the time of inspection (see p. 1 and 6 in [30])
PPP	An output specification is defined with the performance standards to be achieved by the delivery of the service with PPP. They are specified for the control so they should be clearly defined, measurable in quantitative and qualitative terms, and verifiable by the third party (see p. 33 in [31])
FBV	The functional requirements of the road in terms of damage criteria and state condition are predefined in the contract which should be fulfilled by the contractors (see p. 37 in [28]). These indicators are used in comparing and checking the standards of the road during inspection. With FBV performance indicators are included for parts B and C
Conclusion: All these types of contract are performance-based contracts and described by the performance indicators	

Table 4.1 Roles of contractor and owner with FBV, PMMR, and PPP

PBC	Activities								
	Design	Construction	Renovation/ rehabilitation	Structural maintenance	Management	Payment	Operation	Finance	Inspection/ control
FBV					O	O	O	O	O
Part A	O	No	No	No	No	O	No	O	O
Part B	O & C	C	C	No	No	O	No	O	O
Part C	O & C	No	No	C	No	O	No	O	O
PMMR	C	No	C	C	C	O	O	O	O
PPP	O &/or C	C	C	C	C	O or collected from users	C	C	O

Based on analysis of Chap. 3
O owner, *C* contractor, *No* activity not included

6. Key performance indicators (KPI)

PMMR	KPI setting is a difficult task under PRMM because they are the basic of the contracts. Typical performance indicators for roads are International Roughness Index (IRI); absence of potholes, cracks, and rutting; and minimum amount of friction, drainage systems, lane marking, etc. (see p. 7 in [30])
PPP	Performance measures or KPIs are central to the most recent PPP projects which are basic for incentives and penalties; KPIs are, for instance, defined for the lane availability, lane unevenness, potholes, route performance, safety performance, and unplanned events [32]
FBV	The KPI are defined under damage criteria, and state conditions given by ZTV Funktion-StB in Germany (which are KPIs) are allocated for the quality of road under roughness, unevenness, friction, potholes, cracks, ruts, etc. (see pp. 4–6 in [33]). With FBV the KPIs are applied only for contract parts B and C
Conclusion: All the three types of contracts are based on KPI and have similar KPIs for road projects	

7. Payments

PMMR	The payment is based on how well the performance standards are managed on the fixed price lump sum through uniform installments (see p. 3 in [10])
PPP	The payment is based on fulfilled standard criteria and paid by the government to the private sector timely according to the agreement in the contract; in some PPP projects, the payment or part of payments flow from the public users directly (see p. 4 in [34])
FBV	The payment in FBV is through the public fund on meeting the predefined standard either one fixed lump-sum amount or uniform installments; the payment is done after the inspection (see p. 188 in [35])
	Part A: the payment is based on the unit price basic on completion of the project
	Part B: the payment is made after the completion of the construction or renovation (normally 90 % and remaining 10 % after successful results of inspection), but these figures could be changed according to the agreement, for example, with A 93, the 50 % of the amount was paid after construction and the remaining after inspection (see appendix D)
	Part C: the first payment is made in the 9th year and then every 3 years till the expiry of the contract. The payment is made only in the case of fulfillment of standard which is evaluated through inspection
Conclusion: The payment with all three types of contracts is based on the fulfillment of the performance indicators. With PMMR and FBV the payment is from the public sector but with PPP the payment is either through the public sector or collected from users directly	

8. Finance

PMMR	The financing of performance-based contract is through the road agencies from the public fund (see p. 1 in [10])
PPP	The private sector finances the whole project or a part from its fund (see p. 4 in [34])
FBV	The financing of FBV is through the public budget (see p. 188 in [35])
	Parts A, B, and C all are financed through the public fund
Conclusion: Financing with PMMR and FBV flow from the public budget but with PPP financing is through the private fund	

9. Degree of freedom (based on analysis of Chap. 3)

PMMR	The contractor has freedom to be innovative because "what" and "how" the tasks are to be done are not the matter of PMMR but the output should meet the defined standard
PPP	The defined goals, objectives, and standard are to be met, so the contractor has freedom to choose the process and procedures to be innovative
FBV	Performance criteria are to be fulfilled under FBV, so the contractor has freedom to choose the innovative techniques, process, and procedures
	Contract part A is a conventional contract. So the material requirements, process, and techniques are normally defined by the owner. With parts B and C, the contractor has freedom to choose procedures, processes, and material requirements

Conclusion: With all these three types of contracts, the contractor has freedom to be innovative and cost effective. This is however only the principle. Regarding road project, there are requirements and standards like road geometry, color, and pavement thickness. In this sense the contractor has little freedom because they should follow such conditions. Being innovative means accepting risk and risk costs money. To avoid such costs, the contractor probably chooses the techniques and methods that are in practice

10. Special purpose vehicles

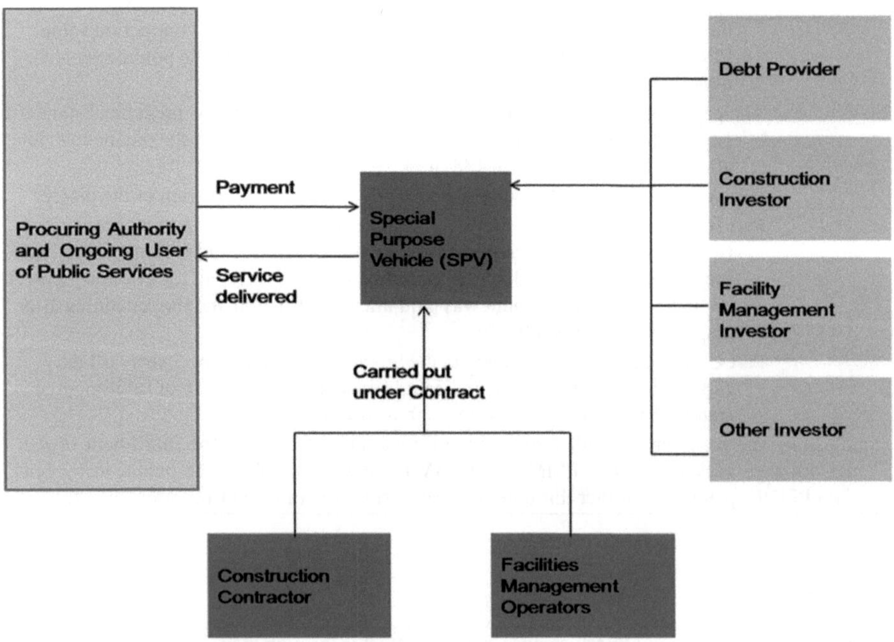

Figure 3.17: typical PPP structure

PMMR	The literatures from the World Bank don't mention PMMR structure with SPV. There is a direct contract between the contractor and the owner. The structure of PMMR doesn't include SPV
PPP	SPV is the feature of PPP. SPV is the contract partner for public sector. SPV is normally established by the private consortium. SPV contracts with client, financial institutions, investors, and other contractors. SPV is limited in nature and helps the private sector in case of bankrupt
FBV	FBV structure doesn't feature SPV. The contractor within FBV is direct between the owner and constructor. FBV projects are normally smaller than PPP and financed by the government. Therefore, the need of SPV has not yet been realized
SPV is the feature of PPP	

11. Phases

PMMR	Plan (work selection) – design – maintenance and management (see p. 7 in [30])
PPP	Initiative (plan) – design – build – operate – transfer (see Fig. 3.19)
FBV	Plan (design) – build – maintain (see p. 207 in [36])
	Part A: includes only initial works like setting construction site and lane marking. So build and maintenance are not included
	Part B: includes renovation or construction of the project. This phase include plan in terms of work selection, staffs selection, and construction of the project
	Part C: includes maintenance task. The necessary plan for the maintenance work is carried
Conclusion: With PMMR there is no provision of new construction till now, whereas with PPP and FBV, construction is performed by the contractor. Operation is included only with PPP, and maintenance work is included with all three options	

12. Risks

PMMR	Work selection, design, delivery, choice, and application of technology, process, and management are carried by the contractor, and hence is allocated higher risks than the owner (see p. 2 in [30])
PPP	Most of the risks are borne by the contractor because he carries out most of the tasks; risks are allocated to that party who is able to manage it properly (see p. 5 in [34])
FBV	Most of the risks are borne by the contractor as contracting period being longer and service type as well as the maintenance work is transferred to the private sector (see p. 37 in [28]) (this applies to contract parts B and C, not for A)
Conclusion: This is in principle. However, the statement is not true for FBV because with part A contract of FBV, the risk is not transferred to the contractor. The detail on risk allocation is shown in Table 4.2	

Force majeure: Risk because of the third party innervations or natural disasters or because of the third party failure (e.g., supplier).

Table 4.2 Comparison of risks with FBV, PMMR, and PPP

PBCs	Risks											
	Design	Construction	Maintenance/ renovation	Quality	Availability	Land acquisition	Force majeure	Political	Inflation	Ground	Traffic volume	Material
FBV												
Part A										No		
Part B	C	C	C	C	C	O	O	S	S	NO	S	C
Part C	C		C	C	C	O	O	S	S	NO	S	C
PMMR	C		C	C	C	O	[a]	O	[a]	NO	O	O
	C	C	C	C	C	O	O or S	O	O	C	S	C

Note: 1. With PPP all risks can be transferred with the provision of economical compensation (Sect. 3.5.7), so allocation may differ in different context
Note 2: Most of the available literature regarding FBV and PMMR include design as the function of the contractor and the design risk is carried by the contractor. However, regarding road projects there is limitation in designing because the properties and structure regarding road are already defined. The aspects like slope, geometry, thickness of each layout, and the corridor are already defined. So there is little room for the contractor in the phase of design. PPP may include freedom in designing projects like schools and prison
C contractor, *O* owner, *S* shared between contractor and owner
[a]Different strategies are applied in different countries (see Sect. 3.3.3 for detail and source)

Traffic volume: In case if traffic volume is higher than estimated, the load on the road is high, which may decrease the life of superstructure or cause extra damage. Because of such defects the contractor needs more maintenance work which would ultimately cause extra money for the contractor. Therefore, this risk is to be compensated by the owner with PMMR and FBV.

PMMR only maintains and manages the road project while FBV has been till now implemented only in road superstructure. In these cases, FBV didn't include ground risks.

Corridor fixing and acquisition tasks cannot be carried by the private sector. Therefore, this risk is handled by the owner.

Force majeure is managed by the contractor in Virginia, USA, while such risks are compensated by the owner in Argentina with PMMR (see p. 3 in [10]).

13. Cost-Effectiveness

PMMR	Reduction of overall costs is achieved through the application of more effective and efficient technologies and work process (see p. 4 in [30]). Reduction in the costs is also achieved from the reduction of overhead costs and staffs because of better management of contracts
PPP	Under PPP, private sector determines the required input to achieve the specified output and apply innovative approach into design and development to reduce overall costs (see p. 5 in [34])
FBV	Cost saving is achieved in the privately constructed and maintained project compared to public operated project; cost minimization is achieved through innovative approach (see p. 38 in [28]). The owner minimizes its costs through synergy by handing over the project to the private sector on long-term basis. The bundling of the construction and maintenance under long-term consideration is more economical because of innovative solution [37]
	Part A is a conventional contract so no point to discuss on cost reduction aspect
	Part B has potential of cost reduction compared to conventional contract but none of the literature mentioned cost reduction is achieved in the road construction or renovation under FBV
	Part C: Since parts B and C are interrelated. If part B is carried out well, cost reduction could be achieved during maintenance in part C
Conclusion: In practice, the report of the World Bank has shown the cost reduction through the implication of PMMR over traditional methods. Such cost reduction has not been reported with FBV. The principle of PPP is to use private finance for public projects. PPP projects suffer the high interest rates and high transactions. So cost reduction with PPP is difficult. As discussed earlier, there is little room to be innovative in the context of road projects because planning (corridor setting) should be carried by the owner and other standard regarding road construction is already defined. Privately operated projects are seen bit cost efficient compared to the public operated. But this amount is too little when compared to the total costs of the project	

14. Quality

PMMR	Improves quality through (based on analysis of Sect. 3.3): Quality is maintained because it is a long-term contract and includes maintenance task The timely inspection is conducted and is the basis for the payments; thus, the quality is always maintained User satisfaction should be ensured for the payment; therefore, quality is always maintained
PPP	Improves quality through (based on analysis of Sect. 3.5): Use of know-how of the private sector To be beneficial during operation, construction and maintenance works is to be qualitative Quality construction cuts the costs during operation and maintenance
FBV	Improves quality through (based on analysis of Sect. 3.4): Functional standard should be met which ensures quality Use of knowledge of the private sector The involvement of private sector from planning and design phase
Conclusion: All the three contracting types ensure quality through similar characteristics. PPP however doesn't include private sector in the planning phase	

15. Area of implementation

PMMR	PMMR has been used in the road rehabilitation, management, and maintenance (see p. 1 and 2 in [10])
PPP	PPP has been used in the various infrastructure projects like roads, railways, water supply, and hospitals (see Sect. 3.5)
FBV	FBV has been developed for road project and exclusively used in the area of superstructure [38]
Conclusion: PMMR includes either only few assets or the whole road assets of an existing corridor and performs the management and maintenance of the assets. Whereas PPP includes wide range of infrastructure and performs the new construction, maintenance, and operation tasks. FBV is developed for the new construction or rehabilitation and maintenance of the road, but it has not been used in the construction of the road but has been used explicitly in the area of superstructure	

16. History

PMMR	The first performance-based road maintenance was started in 1988 in British Columbia, Canada [30] (see p. 2 in [10])
PPP	The first PPP project started in 1992 in form of Private Finance Initiative (PFI) in the UK (see p. 10 in [39])
FBV	The first FBV was piloted in Germany in 2002 (see p. 38 in [28])
Conclusion: PMMR is the oldest form of contract compared to PPP and FBV, but PPP is the most popular and frequently used form of PBC, whereas PMMR and FBV are emerging contracts in the field of road projects	

17. Applied countries

PMMR	Canada, Australia, New Zealand, USA, Argentina, Uruguay, Brazil, Chile, Colombia, Ecuador, Guatemala, Mexico, Peru, UK, Sweden, Finland, Netherlands, Norway, France, Estonia, Serbia Montenegro, South Africa, Zambia, Chad, Philippines, preparation is made to be implemented in Albania, Cape Verde, Madagascar, Tanzania, Burkina Faso, India, Cambodia, Thailand, Indonesia, Vietnam, and Yemen (see p. 2 in [10])
PPP	Since 1990s, PPP has been applied across the world (see p. 19 in [40])
FBV	Germany
Conclusion: PMMR is emerging contracting method in the world, while PPP is already popular and FBV is in the phase of developing	

18. Selection process (explained in detail in Sect. 4.2.1)

PMMR	Based on the "the best value" competitive bidding approach (see p. 3 in [10])
PPP	Unsolicited proposals or direct negotiations ("sole sourcing") or competitive negotiations and competitive bidding (see p. 72 in [29])
FBV	Restricted procedures till now ("Nicht offenes Verfahren mit öffentlichem Teilnahmewettbewerb"). For example, with project A61, restricted procedure was applied (see p. 38 in [28])
	Contract parts A, B, and C are until now not awarded separately
The selection of PMMR is the best value approach (based on past performances, experiences, technical ability, and bid price) and not necessarily the lowest value approach; PPP applies a variety of selection approach while FBV is in the phase of experiment	

19. Costs in terms of interest rates

PMMR	Interest rate for public borrowing is lower than the private borrowing. Financing under PMMR is through the government; thus, project doesn't suffer from the high interest rate of bank which helps eventually to cut the overall costs
PPP	Since the funding comes from the private sector, the PPP projects suffer from the higher interest rate from the bank which may raise the overall project costs
FBV	Funding from the government so FBV projects do not suffer from the higher interest rate of the bank
Conclusion: Only PPP suffers from the higher interest rate of the bank	

4.2.2 Comparative Study of Performance-Based Road Management and Maintenance Contract (PMMR) and Funktionsbauvertrag (FBV)

1. Functional requirements (based on the respective chapters so references are given in the respective chapters)

As Table 3.6, functional requirements with PMMR.

As Fig. 3.12, functional description of state condition with FBV (parts B and C). Also see Fig. 3.9 for damage characteristics.

Damage characteristics define the possibility of defects in different section of the road, while state condition presents the exact features to be considered during inspection.

With PMMR the exact values are defined which should be maintained (IRI < 2 in Argentina or IRI < 2.8 in Uruguay), but with FBV (state condition) each requirement (flatness, rut depth, grip, etc.) is measured with scale (1 (very good) to 5 (very bad)). For the project acceptance the state condition <= 1.9 (it is however not clear if this is the average value of all aspects, but it makes logic that every aspect should be at least under this value, because if grip value is 5 and other aspects are 1, which makes the acceptable value in average, the state condition of road with grip value 5 is not good). So grip, rut depth, cracks, and other features <= 1.9 for project acceptance.

Conclusion: FBV and PMMR have similar features described in the functional requirements. In PMMR each feature is defined with the respective value, but with FBV each feature in general should at least maintain good structural condition which in Germany defined as 1.9 in the scale of 1 (very good) to 5 (very bad). The respective value with PMMR differs from one country to another. This may be because of weather condition or speed limited condition.

Note Source: respective chapters

2. Penalty criteria

Performance standards	Penalty in US$
No pothole >2 cm deep on paved roads	110/day/pothole
No edge failure on paved roads	110/day/failure
No rutting >20 cm long and 12 mm deep on paved roads	66/day/rut
No cracking or raveling on paved roads	88/week/km
Travel speed of at least 50 km/h on earth 176/day/km and 70 km/h on gavel roads	176/day/km
No potholes >2 cm on paved shoulders	44/day/pothole
Drains, ditches, culverts and other drainage structures to be clean	44/day/structure

As Table 5.7, penalties regulation with PMMR on noncompliance or not meeting standards

PMMR has clear rules regarding no compliance or not meeting the functional requirements. The penalty is calculated in terms of money and deducted from the payment. With FBV in such cases:

Part B: Certain percentage of the amount as agreed in the contract is paid after the inspection. If some defects are seen or the standard is lower than defined during inspection, the remaining payment is not done. In this case the contractor is supposed to carry out extra work to maintain the defined standard.

Table 4.3 Evaluation criteria and their weights applied for evaluation of bids in Washington, DC, USA, 2000

Criterion	Weight (%)
Technical (experience, knowledge)	20
Staffing, quality, control/quality assurance, management	15
Past performance	15
Costs (bid price)	50

Stankevich et al. (see p. 5 in [10])
Source: D.C. Department of Public Works

Part C: It is the maintenance contract. In the case of defects during the inspection, the contractor is required to work on to remove all these defects.

With FBV, such penalties in terms of money as with PMMR are not applied. With FBV it should be cleared in how many days the defects are to be maintained and what happens if the contractor takes more than a week or even a month to remove such defects. PMMR has the clear rule which could be adopted by FBV.

3. Inspection

PMMR	Road agency carries the responsibility for monitoring the performance standards. The road agencies contracts experts as supervisors. These supervisors inspect on monthly basis and publish annual report on the performance of the contractor. Throughout the contract period the road should meet or exceed the minimum predefined level (see p. 9 in [41])
FBV	Contract part B: After the completion of the construction or renovation, the inspection is done visually and with the help to measurement techniques. Techniques may differ to measure ruts or cracks or grip, etc.
	Part C: The inspection is conducted by the owner in the rolling basic of every 3 years on state condition and damage characteristics (source Sect. 3.4)

Conclusion: PMMR conducts inspection every month but reports are published every year. With FBV with part B, inspection is done once on the completion and every 3 years with part C

4. Selection of contractor

PMMR	Based on the World Bank document (also see Sect. 3.3), the selection of the contractor is through the "best value approach." Best value approach is selecting the contractor evaluating the contracting company's technical ability, past performances, knowledge, staff quality and qualification, financial ability, and bid price
FBV	With the selection of contractor through restricted procedure ("Nicht offenes Verfahren mit öffentlichem Teilnahmenwettbewerb"), not only the bid price is considered but all the past experiences, past performances, qualification of staffs, and technical ability of the contracting company are evaluated (see p. 1 in [42]). All three contract parts are awarded till now to only one contractor

Conclusion: These statements clearly state that PMMR and FBV follow similar selection procedure. With FBV only limited companies could bid which are familiar to the owner but with PMMR, the bidding is open to all

There is weight for each of the criteria used for selection with PMMR which is not clear with FBV.

Example: Evaluation criteria and their weights applied for evaluation of bids in Washington, DC, USA, 2000 (Table 4.3)

5. Payment condition

PMMR	With PMMR the payment is made on fixed lump sum on installments as agreed in the contract. An example of payment method in Argentina with 5 years road rehabilitation project: Fifty-five percent of the amount was paid in the first year in three installments and remaining 45 % on monthly installments (equal monthly payments on 48 installments) till the expiry of the contract. The payment followed after monthly inspection. It is however not the rule (see p. 10 in [25])
FBV	Part A: Paid after the completion of work
	Part C: First payment after 9 years and then every 3 years
	Part B: Normally 90 % after the completion of construction or renovation, the remaining 10 % after successful results of inspection
	But, with the project A 93, 50 % of the payment was made on completion of the construction and the rest of the 50 % was paid after the inspection (according to Autobahndirektion Süd, see Appendix D)
Conclusion: There are no clear regulations for payment amounts with FBV and PMMR and depends upon the agreement	

6. Unpredictable costs

PMMR	In Virginia, USA, risks for unpredictable costs including inflation, escalating material prices, accidents, and force majeure events are carried by the contractor. In Argentina, 25 % of such costs are reimbursed by the owner. In British Columbia, Canada, the change of prices of material, labor, and fuel is considered by the owner (see p. 5 in [10])
FBV	With FBV unpredictable costs from force majeure events are carried by the owner. A cost through inflation is shared between the owner and contractor as per agreement (see Table 3.10)
Conclusion: FBV and PMMR differ in allocating unpredictable risks. PMMR even differs from one country to the other	

Key Learning of the Section[1]

Comparative Study on Traditional Construction Contracts and Performance-Based Contracts

1. PBCs have been implemented since about 20 years, whereas the traditional contracts have evolved over long time.
2. With traditional contracts the process, procedures, material requirements, and technologies are specified by the owner but not with PBCs. PBCs specify the output requirements of the project.

[1] Based on Sect. 4.2; for the respective references, see Sect. 4.2.

3. Payment under traditional contract is followed after completion of the project, but the payment under PBC is followed only if the completed project meets the specified standards.
4. The selection of contractor under traditional contract is based on the lowest bid approach, whereas the selection of contractor under PBC is normally based on best value approach.
5. Shearing of risks approach is not applied with traditional contract, but with PBC, construction and design risks are borne by the contractor.
6. The duration of traditional construction contracts are normally lower than PBCs.
7. SLA and KPI are not included within traditional contracts whereas they are important components of PBCs.
8. The role of private sector ends with the completion of construction, but with PBC the contractor's role is expanded till maintenance and in some case also operation.

Comparative Study on Performance-Based Road Management and Maintenance Contract (PMMR), Funktionsbauvertrag (FBV), and PPP

9. PPP is for various infrastructure projects while PMMR and FBV cover only road projects.
10. All these contracts are "output" based.
11. All these contracts are long-term contracts and can go up to 30 years.
12. All these contracts are defined with the help of SLA and KPI.
13. Payment with PMMR and FBV is through public fund whereas with PPP, it could be collected directly from the users or through the public fund.
14. Financing with PMMR and FBV is through public fund, whereas the whole or a part with PPP is financed by the private sector.
15. Shearing of risks principle is applied with all these contracts.
16. PMMR and PPP are implemented worldwide, whereas FBV is applied only in Germany.
17. PMMR and PPP select the contractor depending upon competitive selection process but FBV follow limited selection procedure.
18. PPP covers the new road construction/renovation, operation, management, and maintenance; PMMR covers only renovation/maintenance and management of road. FBV covers new road construction/renovation, maintenance, and management, although FBV has been explicitly applied in the area of road superstructure.

References

1. The American Institute of Architects. Architectural graphic standards for residential construction. Hoboken: Wiley; 2010.
2. Zimmermann J. Project delivery systems, Lecture note in Lehrstuhl für Bauprozessmanagement und Immobilienentwicklung an der Technischen Universität München, Issue January 2009.
3. Mincks WR, Hal J. Construction jobsite management. New York: Thomson Delmar Learning; 2004.

4. Competitiveness of the Singapore economy – a strategic perspective. Singapore: Singapore University Press; 1998.
5. Performance – based contracting: here to stay, but challenges ahead. Fall 2004.
6. Hodge G, Greve C. The challenge of Public-Private Partnerships: learning from international experience. Cheltenham: Edward Elgar; 2005.
7. Molenaar KR, Yakowenko G. Alternative project delivery, procurement, and contracting methods for highways. Reston: ASCE; 2007.
8. Newcastle City Council, Australia. www.newcastle.nsw.gov.au/__data/assets/pdf_file/0009/92628/140-A.pdf, p.1. Accessed 11 Apr 2010.
9. Zimmermann J. Project delivery systems, lecture note in Lehrstuhl für Bauprozessmanagement und Immobilienentwicklung an der Technischen Universität Müncen, Issue January 2009.
10. Stankevich N, Qureshi N, Queiroz C. Performance-based contracting for preservation and improvement of road assets. Washington, DC: The World Bank; September 2005 (updated August 2009).
11. Procurement of performance-based management and maintenance of road: the World Bank Document, Washington, DC, Revised March 2003, Preface no.4.
12. Wolper LF. Health care administration, planning, implementation, and managing organized delivery system. Canada; 2004.
13. Kwak YH, Bushey R. Construction management at risk: an innovative project delivery method at storm water treatment area in the Everglades, Florida, ASCE Congress 6.
14. Wearne S. Engineering management – civil engineering contracts. London; 1989.
15. American Water Works Association. Construction contract administration, Denver, USA; 1999.
16. Cameron Jr JG. A practitioner's guide to construction law. Philadelphia: American Law Institute-American Bar Association Committee on Continuing; 2000.
17. Zietlow G. Cutting costs and improving quality through performance specified road maintenance contracts. Pilot experiences in Latin America offer lessons for Africa, Africa Transport Technical Note, July 1998.
18. Shiffmann F. IVT ETH Zurich, Performance-based specification for the operational and structural road maintenance of communal road network. Conference paper STRC 2007, Monte Venta/Ascona, September 2007.
19. University of Reading. Performance-based contracting. www.icrc-reading.org/projects/projectdetails.asp?projectid=13. Accessed 11 Apr 2010.
20. Deloitte Research Study 2006. Closing the infrastructure gap. Role of public-private partnerships. New York: Deloitte; 2006.
21. Manual of water supply practice – M47: construction contract administration. USA: American Water Work Association; 1997.
22. AWWA. Construction contract administration: manual of water supply practices, 1st ed. USA; 1996.
23. Sidney S III, Konrath L. Performance based contracting: a viable contract option? Summary of conference, U.S. and international approaches to performance measurement for transportation systems. Conference proceedings 44, Washington, DC.
24. Friedlander MC, Roberts KM. Design-build – the single entity option. Chicago: SCHIFF HARDIN LLP; January/February 1997.
25. Zietlow G. Implementing performance-based road management and maintenance contracts in developing countries – an instrument of German Technical Cooperation, Eschborn, November 2004.
26. Zietlow, G. www.performance-based-road-contracts.com. Accessed 6 June 2010.
27. Shrestha HR. Sustainable approach to road transport maintenance practice for effective service delivery. Policy paper 24, Kathmandu, September 2006.
28. Nösler I. Pavement design in Rahmen neuer Bauvertragsformen. Brüssel, tis October 2003.
29. ADB. Asian Development Bank: Public-Private Partnership handbook. Philippines.
30. Zietlow G. Cutting costs and improving quality through performance-based road management and maintenance contracts – Latin America and OECD experiences-, Birmingham, April 2004.

31. Iossa, E, Spangnola G, Vallez M. Public-private partnerships. Report by The World Bank, September 2007.
32. Chapter 4: PPP project contract management and operation, U.S. Department of Transport. http://international.fhwa.dot.gov/pubs/pl09010/04.cfm. Accessed 12 Apr 2010
33. Forschungsgesellschaft für Strassen – und Verkehrswesen, Arbeitsgruppe Fahrzeug und Fahrbahn: ZTV Function-StB 01, Koeln; 2001.
34. Public Private Partnership (PPP) guideline: Public-Private Partnership Unit, Prime Minister Department, Putrajaya, Malaysia.
35. Beckers T. Die Realisierung von Projekten nach dem PPP-Ansatz bei Bundesfernstraßen, Dissertion, TU Berlin, Fakultät Wurtschaft & Management, Berlin; 2005.
36. Alfen HW, Leupold A. Public Private Partnerships (PPP) in German public real estate sector. In General real estate – yearbook 2007. The Hague: Europe Real Estate Publisher; 2007.
37. Nösler I. Oberbaubemessung im Rahmen neuer Bauvertragsformen (Funktionsbauvertrag), Brüssel, p. 3. www.nynas.com/upload/docs/Bitumen/NOAH.pdf. Accessed 26 Mar 2010.
38. Schmerbeck R. Pilotprojekte mit Funktionsbauvertragen – Vertragliche Sicherung der Oberflächeneigenschaften, München, p. 69. http://www.vsvi-hessen.de/download/20051102/vsvi2005_11_02schmerbeck.pdf. Accessed 26 Mar 2010.
39. Allen G. The Private Finance Initiative (PFI). Research paper 01/117, UK, December 2001.
40. ADB: facilitating Public-Private Partnership for accelerated infrastructure development in India. Workshop Report, December 2006.
41. Performance-based management and maintenance of roads (PRRM) case study 1: Argentina based on Toolkit PPP in highways, The World Bank.
42. Bekanntmachung der Neufassung der Vergabe- und Vertragsordnung für Bauleistung (VOB) Teile A und B, September 2002, Germany.

Chapter 5
Case Study

5.1 Case Study on Funktionsbauvertrag

Road Project: Rastplatz "Blauer Stein" – Miel on the Highway BAB A 61

5.1.1 Introduction

The section Nordrhein-Westfalen between Rastplatz "Blauer Stein" – Miel on the Highway BAB A 61 is the fifth road project in Germany under Funktionsbauvertrag (see p. 92 in [1]). Before this project the other project was implemented on the federal highways BAB A 61, BAB A 81, BAB A 93, and BAB A31 (as shown in Table 3.11). This road section lies in the German state Northrhein-Westfalen. A 61 built in 1970s is a highway in Germany which connects the boarder of the Netherlands near Venlo in the northwest to the Hockenheim and runs parallel to the A 3 on the opposite side of Rhein [2]. The total length of BAB A 61 is 324 km [3].

A. Gajurel, *Performance-Based Contracts for Road Projects: Comparative Analysis of Different Types*, DOI 10.1007/978-81-322-1302-4_5, © Springer India 2014

Table 5.1 Brief presentation of the project under Funktionsbauvertrag

Length	9.7 km
Duration of contracts	15 years
Contracting period	2008–2023 (see p. 30 in [4])
Contracting type	Maintenance (see p. 30 in [4])
Total costs	19.8 million Euro (see p. 30 in [4])
Purpose	Road rehabilitation [5]
Performances	Removal of the rutting and cracks and skid resistance defect for the smooth flow of traffic and user's satisfaction [5]
Traffic volume	About 56,500/day [5]
Section of construction site [5]	
(1) From Blauer Stein to the south (6 km)	Total length of 6 km which includes complete rehabilitation of one bridge
(2) Rest 2,8 km	Includes two bridges for rehabilitation
Principle to divide site into two sections	To minimize the stress of the driver to drive long construction site
Available lane	Always two lanes in each direction are available for the drivers
Objective of the project	To improve the quality of the construction works and to minimize the traffic disturbances during the road rehabilitation works (see p. 3 in [6])

Majority of the information is derived from Strassen.NRW. www.strassen.nrw.de/service/presse/meldungen/2008/080703-01.html. Accessed 15 Apr 2010

5.1.2 *Brief Presentation of the Project (Table 5.1)*

5.1.3 *Contract Parts in the Project*

For the rehabilitation works of this road section of about 8 km under Funktionsbauvertrag, the contracts are divided into three parts as with every Funktionsbauvertrag. They are as follows[1]:

1. Contract Part A
   The costs allocated for contract part A is eight million Euros and is a conventional contract. The contract includes traffic management, earth works, rehabilitation of roads and drainage system, and the bridge within this section.
2. Contract Part B
   The costs allocated for contract part B is 9.5 million Euros. The contract includes certain standards, features, and performances for the quality of the track, which should be met by the contractor. The process and procedure for the required performance level are to be decided by the contractor itself. The road should be renovated as predefined in the contract meeting the prescribed level of performance.

[1] The informations of Sect. 5.1.3 are derived from Strassen.NRW. www.strassen.nrw.de/service/presse/meldungen/2008/080703-01.html. Accessed 15 Apr 2010.

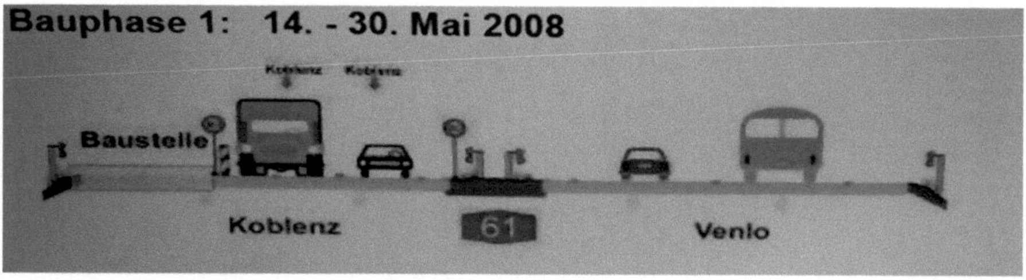

Fig. 5.1 Construction phase 1

3. Contract Part C
   The costs allocated for contract part C is about 2.2 million Euros. The contract includes the preservation and structural maintenance of the rehabilitated road section till the expiry of the contract.

5.1.4 Arrangement of Traffic Route During Construction

For the smooth flow of traffic during the rehabilitation work on road section Rastplatz "Blauer Stein" – Miel of highway A 61, the arrangement was made with changing traffic route. Various consideration was made in order to ensure free flow and without stress to the drivers. The construction site (although it is about 8 km) was divided into two parts, first part was about 6 km and the second part was about 2 km, so that the drivers do not need to cross long construction site continuously. During the construction work, two lanes in each direction were available. The second method was the arrangement of traffic route during the construction [5].

5.1.4.1 Second Method

Rehabilitation of the A 61: In order to have the smooth flow of traffic during the construction, the construction site is arranged in the following way which is described as phases in the first phase (the first 6 km) of construction.[2]

1. Construction phase 1: 14–30 May 2008 (Fig. 5.1)
2. Construction phase 2: 31 May to 01 July 2008 (Fig. 5.2)
3. Construction phase 3: 02 July to 26 August 2008 (Fig. 5.3)
4. Construction phase 4: 27 August to 30 October 2008 (Fig. 5.4)
5. Construction phase 5: 31 October to 14 November 2008 (Fig. 5.5)
6. Construction phase 6: 15–20 November 2008 (Fig. 5.6)

[2] All pictures are based on the Strassen.NRW: www.strassen.nrw.de/service/presse/meldungen/2008/080703-01.html. Accessed 15 Apr 2010.

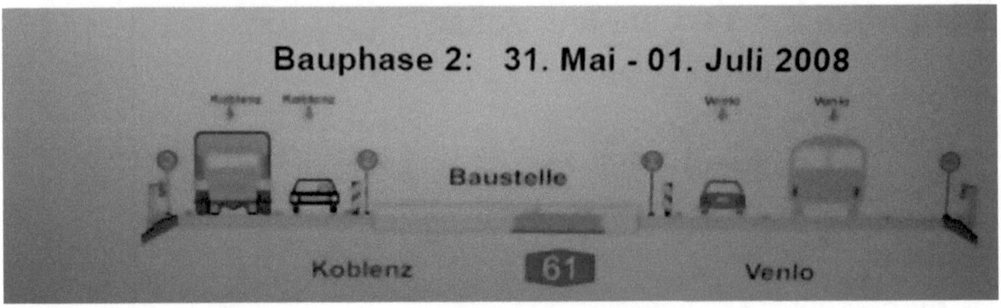

Fig. 5.2 Construction phase 2

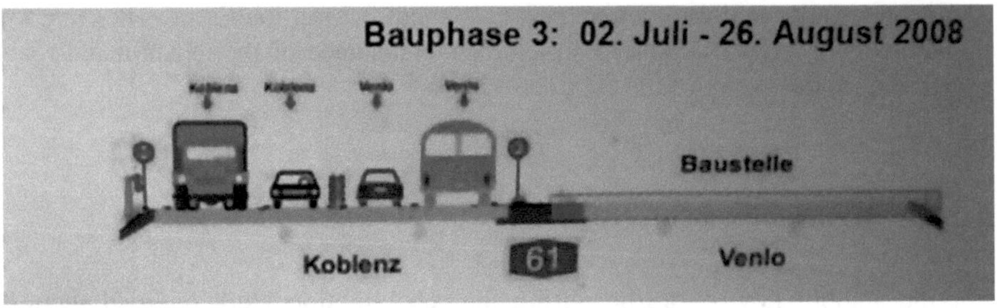

Fig. 5.3 Construction phase 3

Fig. 5.4 Construction phase 4

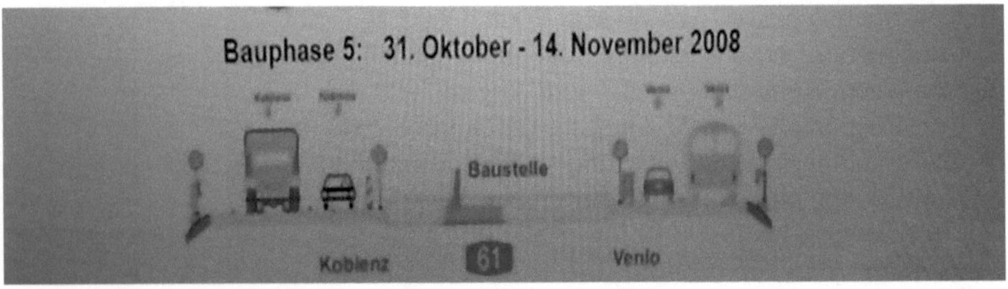

Fig. 5.5 Construction phase 5

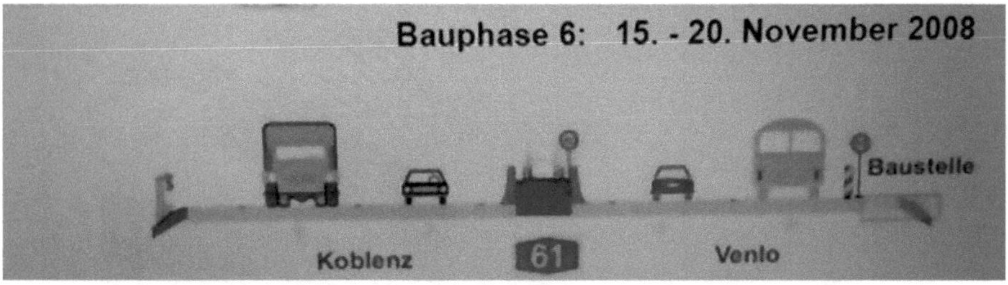

Fig. 5.6 Construction phase 6

With such an arrangement, the free flow of traffic was ensured. According to the Koelnische Rundschau on 17 July 2009, the rehabilitation work of the first 6 km section of the road has been completed. In the second phase, the remaining section would be renovated by the beginning of 2010 [7].

5.1.5 The Contractor

The most economical bidder was selected among the bidders for the rehabilitation and maintenance work. The Koelner Firma Gebrueder von der Wettern GmbH was selected as the most economical and competitive bidder (the selection criteria are not open in literatures). The contractor was not selected only to complete the rehabilitation works working 6 days a week but also to maintain the particular section for the period of 15 years (see p. 3 in [6]).

The contract includes well-defined required quality that has to be met by the contractor. The required levels of performances are defined by the number of technical details. The performance standards are comparable and expressed in the numerical terms. The contractor is required to meet at least the minimum level of performance during the contract period. The safety aspects and the availability of the road section for the free flow of traffic was the main focus of the project. The responsibilities of the contracting company include the inspection of road pavement to ensure absence of ruts and cracks, to maintain the standard grip condition of the road for the proper function of road, and to ensure user satisfaction. In the end of contracting period, the contractor is required to hand over the project with minimum defined quality, which is ensured through the inspection depending upon the technical details.[3] As seen in Fig. 5.7, the road under rehabilitation was done with asphalt pavement.

[3] Main source of information of Sect. 5.1.5 is adopted from Deutsches Baublatt. Issue November 2008, 35. Jahrgang, Nr. 340.

Fig. 5.7 Rehabilitation of the pavement of road under contract [8]

Key Learning of the Section[4]

1. Road Project: Rastplatz "Blauer Stein" – Miel on the Highway BAB A 61 is the fifth project in Germany under FBV.
2. The section lies on highway A 61 in Nordrhein-Westfalen.
3. The length under contract is 8.7 km.
4. The contract covers the road rehabilitation.
5. The total cost of the project is 19.8 million Euro.
6. The traffic volume of this section is about 56,500/day.
7. The contract parts include parts A, B, and C as with every FBV contract.
8. The contracting period is 15 years.
9. The arrangement was made not to disturb the traffic flow during the construction works.
10. The Koelner Firma Gebrueder von der Wettern GmbH is the contractor.

5.2 Case Study on Performance-Based Road Management and Maintenance Contract (PMMR)

Case Study: Performance-Based Road Management and Maintenance Contract (PMMR) in Argentina

[4] Based on Sect. 5.1, the respective references are presented at the end of this chapter.

5.2.1 Introduction

Argentina is the second largest countries in South America after Brazil with GDP per capita of $14,408 [9] and bordered with Bolivia, Paraguay, Chile, Uruguay, and Brazil. Argentina has well-developed road network (also can be seen in Fig. 5.8, Tables 5.2, 5.3, and 5.4) of about 500,000 km, which comprises (see p. 14 in [10]):

1. National road network (Table 5.2)
2. Provincial network (Table 5.3)
3. Municipal network (Table 5.4 and Fig. 5.8)

The concession contracts in the road projects started in 1991, which is being popular day to day. Shortly after the implementation of the performance-based contracts in British Colombia in Canada in 1988, Argentina concessioned approximately 10,000 km of its main road under performance-based contracts; in mid-1990 another 10,000 km was concessioned under such arrangement, and in Argentina such contracts are referred as CREMA which includes the rehabilitation and maintenance of the road (see p. 2 in [13]). According to Liautaud (2004), in 1995 Argentina introduced the performance-based road contract for management and maintenance (see p. 2 in [14]). Such contracts are very much advantageous to the poor countries like Argentina for the preservation of the road assets, which are suffering because of the inadequate maintenance and management of the roads (see p. 155 in [15]). Approximately 12,000 km of road (40 % of national paved road network) in Argentina is rehabilitated under PMMR (see p. 128 in [11]). Argentina is currently working to make preparation to expand performance-based contracts from national to the provincial roads (see p. 6 in [14]). The average cost under CREMA is US$11,000/km/year (see p. 96 in [16]).

5.2.2 Reasons for Implementing PMMR in Argentina

Argentina has well-developed road network; now there is necessity of management and maintenance. This is the one reason of PMMR being popular as it is developed for the maintenance and management of the roads. The other reasons are (see p. 4 in [17]):

- Reduction of the staff to measure and monitor activities
- The need to focus on the user's satisfaction
- The need to transfer the responsibilities and risks to the contractors
- Reduction of transaction costs and other overhead costs
- To improve overall condition of the road
- To improve control and quality standards
- Reduction of the maintenance costs through the innovative approaches

Fig. 5.8 The road network in Argentina (*Red caminera* – Road network, *Autopista* – Highway, *Ruta pavimentada* – Paved road, *Camino consolidado* – Consolidated road, *Ruta national* – National route, *Ruta provincial* – Provincial route) [12]

Table 5.2 National network in Argentina (see p. 129 in [11])

Total length	38,744 km
Paved	30,912 km (80 %)
Unpaved	7,832 km (20 %)

Table 5.3 Provincial network in Argentina (see p. 14 in [10])

Total length	181,000 km
Paved	32,580 km (18 %)
Unpaved	148,420 km (82 %)

Table 5.4 Municipal network in Argentina (see p. 14 in [10])

Total length	280,256 km
Paved	81,274 km (29 %)
Unpaved	19,898 km (71 %)

5.2.3 *The Projects*

Road maintenance and rehabilitation have been traditionally carried with input-based contracts with the private sector, but in order to cut costs and to use innovative approach, output-based construction contract was developed in Argentina. This contract is applied at national level with national highways, initial phase was only for the maintenance contract with scheduled payment based on km/month, and the next phase included maintenance with rehabilitation with lump-sum payment up-front (see p. 40 in [18]).

Since the introduction of output-based contract in 1995, the contract was renewed in 4 years of its application, now covers 3,600 km of paved roads with 11 contracts and the contractor are paid monthly on equal installments on meeting the predefined performance standards and penalties on daily basis if the performance criteria are not met (see p. 40 in [18]).

The total costs of these 11 projects amount to US$650 million and paid in lump sum on monthly basis based on per km/month. The contract was initially for 2 years with the option of renewal. Three inspections were held per month; the payment was followed for meeting the specified standards and penalties for noncompliance or not meeting the specified standards. The average costs of maintenance was US$175 km/month and the penalties that resulted from these entire project was US$300,000, which is just 1 % of the total costs (see p. 32 in [19]).

5.2.3.1 Phases of CREMA

1. *First Phase*

   The km/month contract was developed by the innovative contract which included maintenance and rehabilitation works and called as CREMA. Under CREMA first phase, the 11,818 km of non-concessioned roads for 5 years contract based on lump-sum amount was contracted, and the project ranges from 100 to 300 km with

Table 5.5 Rehabilitation and maintenance conditions and indicators with PMMR in Argentina (see p. 43 in [18])

Rehabilitation and maintenance conditions and indicators
Throughout the contract period rehabilitation works must:
Meet or exceed the minimum predefined thickness of overlay
Not exceed the maximum level of roughness, rut depth, cracking, or raveling
Regular visual inspections of maintenance work focus on a few essential items in ensuring compliance with the specification:
Potholes, cracking, and rutting
The condition of shoulders, culverts and drains, and roadside environment
Guardrails and vertical and horizontal signs

contracting period exceeding 10 years (see p. 32 in [19]). Approximately, half of the payment was made during the first year which covered the rehabilitation costs, and the rest was paid on a monthly basis, penalties were allocated for noncompliance on time and not meeting the standards, and the minimum overlay thickness was specified with the indicators for the potholes, cracks, rutting, and blockage of the drainage system, friction, and deflection (see p. 32 in [19]).

2. *Second Phase*

   It included 8,200 km of non-concessioned roads for the rehabilitation and maintenance with total costs of US$550 million. The road condition of second phase was worse than the first phase but the technical specification of both phases were the same. For the second phase the overlay was thicker as defined by the World Bank HDM model. According to Stankevich (2006) the rehabilitation occurring period was increased by 12–24 months or more (see p. 32 in [19]).

   In another context to finance the road rehabilitation and maintenance, the government awarded concession where the contractor collected the revenues through toll, and the other additional costs not covered from toll collection was paid by the government according to [20] (see p. 32 in [19]).

   Nine thousand five hundred and eight kilometers of national roads are contracted under PMMR, where payment is through the toll collection by the concessionaire with the contracting period of 12 years. The specification of the indicators and penalties for noncompliance were defined clearly. Another 1,897 km of national roads is constructed under PMMR with the duration of 10 years without tolling system. The toll was not possible because of the insufficient traffic; in such case the payment is through the government.

5.2.3.2 Inspection

Road agency is responsible for monitoring the performance standards. For the inspection the supervision department of road agency contracts the supervisors, who inspect on the monthly basis and publish annual statistics on the performance of the contractor (Table 5.5; see p. 9 in [21]).

5.2.3.3 Performance Standards for Pavement (See p. 8 in [21])

Case 1: Bituminous Pavement
Roughness (IRI) <2, 0 m/km
Rutting <12 MM
Cracking: Type 4 (according to the catalogue of the D.N.V.) and less than 15 % of the area evaluated

Case 2: Concrete Pavement
Roughness (IRI) <3, 3 m/km
Cracking: Type 2

Case 3: Bituminous Surface Treatment
Roughness (IRI) <2, 9 m/km
Cracking: Type 4

These are the standard fixed by the road agencies for the contractors. Payment is followed only meeting these standards. Argentina is now considering introducing several new performance indicators for the second phase of CREMA from its experiences with the previous projects (see p. 7 in [14]).

CREMA (PMMR is referred as CREMA in Argentina)

5.2.4 Benefits of PMMR (CREMA) in Argentina

1. *Risk Allocation*
   As with every performance-based contract, most of the risks (design and construction) are transferred to the contractors. In Argentina, in certain condition the reimbursement of cost overruns are made by the owner. For instance, in case if the cost overruns because of the circumstances which are not manageable by the contractor, for example, earthquake, flood, or shortage of some constructing materials. The contractor's schedule of input prices submitted in the bid is the basis for cost overruns estimation. Only 25 % of excessive cost runs is covered by the government (see p. 6 in [14]). Such provision is beneficial to the contractor as it provides some compensation in case cost overruns occurred due to unavoidable conditions and also increases the level of confidence of the contractor for choosing new techniques, procedures, and technologies.
2. *Improved Road Condition*
   Argentina significantly improved the condition of roads implementing performance-based approach. According to Liautaud (2004) it reduced the poor road condition from 25 % to less than 5 % till the end of 1999 (see p. 6 in [14]).
3. *Continuous Supply of the Fund for Road Maintenance*
   PMMR is a long-term contract, so the government is required to allocate the long-term budget. For making long-term payment, the government is required to allocate the budget for road maintenance every year. This obligation helped the continuous allocation of the budget for the road maintenance (see p. 6 in [14]).

Table 5.6 Features of PMMR (CREMA) in Argentina (see p. 96 in [16])

Features	Description
Activities	The contract combines: (1) Initial works (rehabilitation of roads, signaling, and surroundings) (2) The road maintenance activities over the duration of the contract
Payment	The contractor is paid a fixed amount; usually 45 % of the total costs during the first year of the contract for the execution of initial works and the rest 55 % in equal monthly payments till the expiry of the contract
Quality standards	For the paved roads, International Rough Index (IRI) below preestablished ceiling, no potholes allowed, maximum times established to correct cracks and other problems, and execution of agreed maintenance activities, signals, drainage, and surroundings
Award	Competitive bidding procedure and selection based on lowest price bidder
Road agency role	Delegation of operational responsibilities to the contractor including planning and execution of works and selection of technical applications Role of road agency shifts to conceptual (and sometimes detailed) design of project, supervision, and regulating of contracts

4. *Reduction of Risk of Unsatisfactory Quality*
   Inclusion of rehabilitation work with maintenance contract resulted in several benefits because it encourages the contractor to undertake rehabilitation work at higher level to reduce costs during the maintenance phase. According to Liautaud (2004), Argentina has reduced the risk of unsatisfactory quality through the inclusion of rehabilitation with performance-based contracts (see p. 6 in [14]).

5.2.5 Features of PMMR (CREMA) in Argentina (Table 5.6)

5.2.6 Reward and Penalties Criteria

The regular payments are made timely on the achievement of a specified level of services not on the quantities and unit rates. The payment is reward for the contractor. The performance is assessed during monthly inspections carried by the engineers, supervisors, and the contractor. The rehabilitation work must fulfill the maximum level of roughness throughout the contracting period; in Argentina the maximum roughness level is usually 3.3.[5]

Potholes, cracking, rutting on the pavement, condition of shoulders, culverts, drainage system, roadside, vertical and horizontal signs, lane marking, grass on the shoulders, and guardrails are inspected during the inspection to ensure the compliance

[5] The Sect. 5.2.5 is based on Presentation of Performance-Based Management and Maintenance of Road, Cases Study1: Argentina (Source: Toolkit PPP in Highways, The World Bank).

Table 5.7 Examples of penalties for noncompliance of deficiencies [21]

Performance Standards	Penalty in US$
No pothole >2 cm deep on paved roads	110/day/pothole
No edge failure on paved roads	110/day/failure
No rutting >20 cm long and 12 mm deep on paved roads	66/day/rut
No cracking or raveling on paved roads	88/week/km
Travel speed of at least 50 km/h on earth 176/day/km and 70 km/h on gravel roads	176/day/km
No potholes >2 cm on paved shoulders	44/day/pothole
Drains, ditches, culverts and other drainage structures to be clean	44/day/structure

of the specifications. All of these aspects are assigned with the performance standard, e.g., roughness level should be at least 3.3.

The compliance of the required level is followed by the payments, while the noncompliance or deficiencies are followed by reduction in the payment. Examples of penalties for noncompliance of deficiencies with response time with performance-based contracts (CREMA) in Argentina are presented in Table 5.7.

Key Learning of the Section[6]

1. Argentina has a well-developed road network of about 500,000 km.
2. Argentina introduced PMMR in 1995.
3. Approximately 40 % of the national paved roads in Argentina are rehabilitated under PMMR.
4. PMMR is called as CREMA in Argentina.
5. Payment is based on agreed fixed price.
6. The average cost of rehabilitation is US$175 km/month.
7. CREMA defines the SLA and KPI for the performance measures.
8. The selection of the contractor is based on the competitive bidding procedure.
9. Penalty criteria are allocated for noncompliance or not meeting the specified standards.

References

1. Beckers T, Hirschhausen CV, Klatt JP. Reformbedarf bei den Bundesstrassen und das potential des PPP-Ansatzes, Berlin, May 2009.
2. http://citizendia.org/Bundesautobahn_61. Accessed 15 Apr 2010.
3. www.autobahn-online.de/a61.html. Accessed 15 Apr 2010.
4. Geschäftsbericht 2007. Ministerium für Bauen und Verkehr des Landes Nordrhein-Westfalen, Strassen.NRW.
5. Strassen.NRW. www.strassen.nrw.de/service/presse/meldungen/2008/080703-01.html. Accessed 15 Apr 2010.

[6] Based on Sect. 5.2, the respective references are presented at the end of this chapter.

6. Deutsches Baublatt. Issue November 2008, 35. Jahrgang, Nr. 340.
7. www.rundschau-online.de/servlet/OriginalContentServer?pagename=ksta/ksArtikel/Druckfassung&aid=1246895308791. Accessed 18 Apr 2010.
8. www.general-anzeiger-bonn.de/index.php?k=loka&itemid=10001&detailid=469683. Accessed 18 Apr 2010.
9. www.embassyworld.com/maps/Maps_Of_Argentina/index.php. Accessed 18 Apr 2010.
10. Estache A, Carbajo JC, de Rus G. Argentina's transport privatization and re-regulation: ups and downs of a decade long daring experience. 2nd ed. Washington, DC: The World Bank; 1999.
11. Cabana G, Liautaud G, Faiz A. Areawide performance-based rehabilitation and maintenance contracts for low – volume roads. Transportation research record 1625. New Delhi: Central Road Research Institute.
12. www.reservas.net/alojamiento_hoteles/imagenes/argentina_mapaplanorutas_550x800.gif. Accessed 18 Apr 2010.
13. Zietlow G. Cutting costs and improving quality through performance-based road management and maintenance contracts – Latin America and OECD experiences-, Birmingham, April 2004.
14. Stankevich N, Qureshi N, Queiroz C. Performance-based contracting for preservation and improvement of road assets. Washington, DC: The World Bank; September 2005 (updated August 2009).
15. Charlier F, Cho-Oguie C. Sustainable reforms for inclusive growth in Cameroon – a development policy review. Washington, DC: The World Bank; 2009.
16. Burki SJ, Perry G, Dillinger W. Beyond the center: decentralizing the state. World Bank Latin American and Caribbean Studies. Washington, DC: World Bank; 1999.
17. Performance-based management and maintenance of roads (PRRM) Case study 1: Argentina based on Toolkit PPP in highways, The World Bank; Zietlow G. Cutting costs and improving quality through performance-based road management and maintenance contracts – Latin America and OECD experiences-, Birmingham, April 2004.
18. Liautaud G. Maintaining roads – experience with output-based contracts in Argentina. The World Bank Report 2004.
19. NCHRP Synbook 389. Performance-based contracting for maintenance. Washington, DC; 2009.
20. Segal FG. Contracting for road and highway maintenance. Los Angeles: Reason Pubic Policy Institute Publication; 2003.
21. Performance-based management and maintenance of roads (PRRM) Case study 1: Argentina based on Toolkit PPP in highways, The World Bank.

Chapter 6
Conclusion

6.1 Key Ideas

The importance of infrastructure has been realized because it enhances economic as well as social activities. The presence of infrastructure alone however doesn't insure the development, but its absence certainly hampers economic growth and social activities. This statement can be justified with a small example; for instance, if a country doesn't have sufficient transport infrastructure, how would it transport finished and raw materials from one region to the other? Trade cannot take place without developed transport infrastructure. How would the people get access to hospitals, schools, and colleges without proper transportation systems because hospitals, schools, and colleges, etc., are not found in every society? With this argument, it can be argued that infrastructure is a catalyst which helps boost the economic as well as social activities.

Infrastructure development is time consuming and expensive. In the real world, traditional construction contracts are still in implication with performance-based construction contracts. Performance-based construction contracts include maintenance task. Inclusion of maintenance task make the performance-based contract a long term because maintenance is carried periodically and real maintenance work starts some years after the operation. Results of performance-based contracts are positive, but they are not effective for the smaller projects (for instance, construction of 2 km highway with PBC is not meaningful; therefore, in Germany for the implementation of function-based contract Funktionsbauvertrag, the road section should be at least 10 km long). Thus, traditional construction contracts would be implemented with the performance-based contracts in the infrastructure development. The traditional construction contracts significantly differ from the performance-based contracts because they are "input" based while performance-based contracts are "output" based. The material requirements, procedures, and techniques are normally defined by the owner in traditional contracts, but the contractor has the freedom to choose the material requirements, procedures, and techniques with performance-based contracts.

A. Gajurel, *Performance-Based Contracts for Road Projects: Comparative Analysis of Different Types*, DOI 10.1007/978-81-322-1302-4_6, © Springer India 2014

From the literature review and the common practice, public-private partnerships (PPP), performance-based road management, maintenance contracts (PMMR), and Funktionsbauvertrag (FBV) are commonly used performance-based contracts for project delivery. PMMR and FBV are developed for the road project delivery whereas PPP comprises different infrastructures like roads, hospitals, water supply, and railway among others. However, principally FBV and PMMR could be developed for other project delivery by developing the clear functional requirements for each type of projects. PMMR has shown a cost reduction of 10–40 % over conventional contracts, as presented by the World Bank document (see p. 6 in [1]). FBV is developed and implemented in Germany and its initial results of so far implemented projects are positive.[1] PPP is growing popularity which could be realized by the number of projects implemented under PPP worldwide.

PPP, PMMR, and FBV are all output-based contracts where the owner defines the output specifications and not the material requirements, processes, and procedures. The contractor has the freedom to choose the techniques, procedures, and methods to deliver the project. The payment is based on meetings or exceeding the specified standards. The analysis of these contracts has shown that they are similar to each other in principle but differs in some aspects such as process, procedures, risk allocation, finance, operation, payments, role of contractor and owner, and other legal arrangement like in the selection procedure. Principally they are similar but not identical; each has its own features and characteristics. Funktionsbauvertrag is further divided into three contract parts: part A (is like a conventional contract and includes initial works before construction), part B (is defined with functional requirements and includes maintenance or renovation works), and part C (is defined with functional requirements and includes maintenance work). Such contract parts are not included with PPP and PMMR.

The analytical study has shown that PMMR and FBV have similar features and characteristics compared to PPP. Firstly, PMMR and FBV are till today applied only for road projects while PPP includes roads projects with others. Secondly, payment and finances with PMMR and FBV are through the public fund but financing with PPP is normally through the private fund and financing is either through the government or collected directly from the users. Thirdly, operation of the project is carried out by the private sector with PPP, whereas the operation is carried out by the owner with FBV and PMMR.

The literature on PPP, FBV, and PMMR includes design as the task of contractor. In case of PPP, it could be applied in the construction of building, prison, or school. But in case of the road project, the corridor is planned by the public sector and the design of road regarding cross section value, gradient, drainage system, inner and outer layer thickness, geometry of the road, color of the superstructure, etc., are already presented in the design manual of each country, which should be fulfilled by the contractor. Therefore, the contractor has little chance of involving in designing the project. After the design phase is complete, the contractor could design the project in detail before constructing, work selection and select the material requirements,

[1] http://www.strassen.nrw.de/service/presse/meldungen/2008/080703-01.html, Accessed on: 10.07.2010.

techniques, staffs, and procedures. In this sense the contractor of PPP, FBV, and PMMR has little effect on designing the road project.

The next difference is that with PPP road project (detail design), construction, finance, operation, and maintenance are carried by the private sector, whereas finance and operation are responsibilities of the owner under PMMR and FBV. Responsibilities of the contractor are expanded with PPP compared to PMMR and FBV because it includes financing and operation aspects.

The noticeable difference between PMMR and FBV is that under FBV the new road construction, renovation, and road maintenance are included, whereas under PMMR new road construction is not included; it is developed only to renovate, maintain, and manage the road. They also differ significantly in terms of selection of the contractor; PMMR selects the contractor through best value approach with a competitive approach where as in practice FBV selects through a restricted procedure ("Nicht offenes Verfahren mit öffentlichem Teilnahmewettbewerb"). FBV also includes the best value approach (past performances and experiences, technical and financial ability, qualification of staffs, and bid price); however, PMMR follows open selection procedure, while FBV follows restricted procedures (only selected companies could apply for bidding). FBV and PMMR also differ in terms of management of the project. With PMMR, the private sector manages the road project, while the owner manages the road project under FBV.

Under PPP, the private sector normally finances the public infrastructure project through the loans from financial institution. This provision takes the responsibilities of the government to finance the public infrastructure through the public fund. The private sector collects its investment directly through the users in some cases (e.g., road project through road tolls) or a part is collected through the users and remaining is paid by the government or the whole is paid by the government on the completion of the project on meeting the specified standards with PPP. Under FBV and PMMR the finance and the payment are carried out by the public part. Thus, the PPP projects suffer the high overall costs compared to FBV and PMMR because the financial institution allocated higher interest rates on private borrowing than on public borrowing. In this sense, PPP projects are costly compared to FBV and PMMR because they suffer high interest rates of financial institution because the borrowing is done by the private part. Regarding duration, they are all long-term contracts and go up to 30 years and are defined with the key performance indicators and service level agreements, which should be met by the output.

Allocation of the risks is another factor which determines the effectiveness of project delivery methods. Risks add extra costs to the projects. Therefore, the contractor as well as the owner looks for the delivery method that manages the risk more efficiently, which eventually cuts the project costs. The principle of risks allocation with PPP, FBV (contract parts B and C), and PMMR is similar. Transferring the risk among the partners who can best manage them with minimum costs is the principle of these three delivery methods as presented by the analytical study. However they differ in some aspects, for example, material availability risk is borne by the contractor with FBV and PPP, whereas it is handled by the owner with PMMR. But the main risks: construction, design (where the contractor is involved), and maintenance remain with the contractor with all three project delivery methods.

The case study of FBV (Road Project: Rastplatz “Blauer Stein”- Miel on the Highway BAB A 61) presented the practical example on how FBV has been implemented. It has supported the analytical study of Funktionsbauvertrag by presenting the real world example by providing detailed data on the length of the section, duration of contracts, costs, objectives of the project, purpose, construction arrangement, contract parts, responsibilities of the contractor, and carried performances. The other case study of PMMR (Performance-Based Road Management and Maintenance Contract in Argentina) introduced the road network of Argentina, which is well developed. The case study supported the analytical study of PMMR with the experience of Argentina. PMMR which is known as CREMA is very popular in Argentina. About 40 % of the national paved roads are maintained under PMMR. The average cost of road maintenance is around US$175 km/ month. The other important aspects such as selection procedures, the performance standards, benefits, risk allocation, finance and payment, features, and reward and penalty criteria of the real world implications are presented. The real world experiences have fit in well with the analytical studies. For instance, the analytical study stated that noncompliance of the stated performance results in the penalties for the contractor which is also applied in the real world experiences as shown by the case study of Argentina (Table 5.4).

6.2 Recommendations

The book presented the analytical study of performance-based road management and maintenance (PMMR), Funktionsbauvertrag (FBV), and PPP. The analytical study leaded to the comparative study between them. Some aspects couldn’t be presented due to time limitation and scope of the book; the following aspects could be research subjects:

(a) The positive experiences of PPP and FBV have been realized. The financing aspects of PPP through the private partner could be combined with FBV so that the public fund can be used on other public sector like health care, education, and poverty reduction programs, etc.
(b) Literatures regarding performance-based road management and maintenance contract and Funktionsbauvertrag state that the design of the project is carried by the contractor. What exactly the contractor deigns is not clear.
(c) FBV has unique contract parts (A, B, and C) which have simplified the contract to implement.
(d) FBV and PMMR are similar in principle. PMMR has long history compared to FBV. Therefore, FBV could adopt the successful aspects of PMMR like the “best value approach” selection procedure and clear penalty criteria.

(e) Special purpose vehicle (SPV) plays a vital role within PPP acting as a contact partner for government, banks, and subcontractors. It helps the private sector in case of insolvency because it is limited in nature. It also helps in raising fund for project investment. The structure of FBV and PMMR do not include SPV.
(f) Maintenance and management tasks would be efficient when combined with operation. So, FBV and PMMR may be more efficient combined with operations along maintenance and management.

Reference

1. Stankevich N, Qureshi N, Queiroz C. Performance-based contracting for preservation and improvement of road assets, September 2005 (updated August 2009). Washington, DC: The World Bank; 2005.

Appendices

Appendix A. Interview Protocol with Dr. Wolfgang Wüst and Mr. Fischer

Location of interview:	Autobahndirektion Südbayern, München
Time:	01 July 2010, 10:00 to 12:00
Protocol taker:	Mr. Benno Vocke
Teilnehmer:	Dr. Wüst, Autobahndirektion Südbayern Mr. Fischer, Autobahndirektion Südbayern Mr. Vocke, Research Associate, TU München Mr. Gajurel, Student TU München
Distributor:	All participants

Nr.	*Protocol points*
	Roads are the basic needs of mobility. The human interactions as well as economic activities are dependent on how well road infrastructures are developed. On the one hand, road construction is expensive, and on the other hand, it takes much time. But the governments of developed and developing countries have limited budget for road infrastructure development because the public fund should cover all the expenses and investments of the state. The public funds should cover sectors like health, education, agriculture, poverty reduction, power supply, and water supply. To overcome such financing and time constraint, various road project delivery methods (e.g., design-bid-build and design-build) are developed. With time, other delivery methods are evolved with eliminating the disadvantages of the previous delivery methods and with new processes, procedures, and principles for effective and efficient project delivery. In Germany, Funktionsbauvertrag has been developed for the road project delivery. It has been developed with the aim to reduce the costs with quality improvements by expanding the role of private sector (design (plan), construction, and maintenance) and financing through the public fund

(continued)

(continued)

1	*Funktionsbauverträge in general*
1.1	*What are the reasons in your opinion for and against the choice of a Funktionsbauverträge in comparison to conventional procurement?*
	Funktionsbauvertrag is a long-term contract which includes the maintenance task compared to the conventional procurement. This property ensures better quality construction compared with conventional. Most of the risks are transferred to the contractor including maintenance risk; even the "Grundbaurisiko" could be transferred to the contractor with some options. Such properties of FBV help in the quality construction. Therefore, FBV is regarded better option than conventional procurement
	What are the reasons in your opinion for and against the choice of a Funktionsbauverträge in comparison to the procurement under a PPP model?
	With PPP, operation of the project, structural maintenance, and the operational management are carried by the contractor, while only structural maintenance is performed by the owner with Funktionsbauvertrag. With Funktionsbauvertrag, it is difficult to distinguish the effect of structural maintenance to the operation. And sometimes, the poor operational management may result defect on roads which should be maintained by the contractor. For example, removal of snow on the road falls under operational management. If the snow is not removed timely, it would cause damage on the surface of the road. This should be now maintained from the cost of contractor under Funktionsbauvertrag. There is a need of extra calculation method to overcome such disadvantage in Funktionsbauvertrag
1.2	*Which project characteristics speak for or against the application of a Funktionsbauverträge?*
	There are no clear characteristics that speak for or against the application of Funktionsbauvertrag. But it is evaluated depending on the investment volume and risk allocation plan. Funktionsbauvertrag is attractive for projects to be contracted for long term. The large- and medium-size projects are suitable for Funktionsbauvertrag
1.3	*For PPP projects, there is feasibility test in Bavaria. Is there a similar document for Funktionsbauverträge?*
	Funktionsbauverträge have been limited to the pilot projects. Therefore, there is no separate document for FBV for feasibility test, but it is in the process of discussion. In selecting the project for FBV, the norms are implemented which already exist in road project delivery. For example, there is condition regarding the minimum length of the road for Funktionsbauvertrag
2	*Development and procurement*
2.1	*After first two pilot projects (A 81 in Baden-Württemberg: Oberndorf-Rottweil; and A 61 in Rhineland-Pfalz: Koblenz-Kruft) in 2002, only more three projects under Funktionsbauvertrag were implemented, and three more are on consideration until 2010; how do you see this development?*
	The development is slow. I see the problem with financing for the slow development of Funktionsbauverträge. In today's context, we cannot expect high number of project under Funktionsbauvertrag. And even we cannot expect much from Funktionsbauvertrag. I hope in the future the government will bring quota system like all 15–20 % of all road projects in Germany should be delivered under Funktionsbauvertrag (personal opinion)
2.2	*Do you consider it meaningful to implement and award contract parts A, B, and C separately?*
	Funktionsbauvertrag is divided into parts A, B, and C, with their unique characteristics. For example, part A is like conventional contract which includes works like lane marking and removing grass, and part B includes functional requirements and includes construction or renovation task, whereas part C includes maintenance work. They are structured in such a way that the task with part A has its impact with parts B and C and the task of part B has effect on part C. That is why they cannot be implemented and awarded separately.

(continued)

(continued)

3	*Penalty and compensation*
3.2	*How is penalty criteria formulated? Are there clear rules of punishment?* There are bonus and penalty criteria with Funktionsbauvertrag. For noncompliance or not meeting standards, the contractor payment is deducted, or he should perform additional tasks to meet the defined standards
3.3	*How do you assess the freedom of private partner regarding designing and construction for the possible cost optimization?* Innovation is limited with Funktionsbauvertrag. The road has its own standard regarding geometry, slope, and material requirements which the contractor should fulfill. Therefore, there is very little room of freedom to be innovative. However, the contractor can choose either concrete or asphalt pavement depending upon the duration of the contractor *End at* 12:00 h *Note: Interview is translated from German to English by Ashish Gajurel (the author of the book)*

Appendix B. Interview Protocol with Mr. Rupert Schmerbeck

Location of interview:	Autobahndirektion Südbayern, München
Time:	01 July 2010, 10:00 to 12:00
Protocol taker:	Ashish Gajurel
Teilnehmer:	Mr. Schmerbeck, Autobahndirektion Südbayern Herr Gajurel, Student TU München

Nr.	*Protocol points*
	Roads are the basic needs of mobility. The human interactions as well as economic activities are dependent on how well road infrastructures are developed. On the one hand, road construction is expensive, and on the other hand, it takes much time. But the governments of developed and developing countries have limited budget for road infrastructure development because the public fund should cover all the expenses and investments of the state. The public funds should cover sectors like health, education, agriculture, poverty reduction, power supply, and water supply. To overcome such financing and time constraint, various road project delivery methods (e.g., design-bid-build and design-build) are developed. With time, other delivery methods are evolved with eliminating the disadvantages of the previous delivery methods and with new process, procedures, and principles for effective and efficient project delivery. In Germany, Funktionsbauvertrag has been developed for the road project delivery. It has been developed with the aim to reduce the costs with quality improvements by expanding the role of private sector (design (plan), construction, and maintenance) and financing through the public fund.
1	*Funktionsbauverträge in general*
1.1	*What are the reasons in your opinion for the choice of a Funktionsbauverträge in comparison to conventional procurement?* Funktionsbauvertrag is a long-term contract which includes the maintenance task compared to the conventional procurement. In Funktionsbauvertrag, it is clear what should be done during the maintenance period

(continued)

(continued)

	What are the reasons in your opinion for the choice of a Funktionsbauverträge in comparison to the procurement under a PPP model? PPP is suitable for the large projects, while Funktionsbauvertrag can be implemented for the large projects as well as for the medium-size projects, but it is preferable for medium-size projects. PPP needs much preparation with higher preparation costs compared to Funktionsbauvertrag. Funktionsbauvertrag has lower tax compared to PPP because of the lower number of transaction and only fewer risks are transferred to the contractor compared to PPP (e.g., financial risk is not borne by contractor with Funktionsbauvertrag). Funktionsbauvertrag is simpler to execute compared to PPP
1.2	*Which project characteristics speak for or against the application of a Funktionsbauverträge?* Funktionsbauvertrag is suitable for the large projects but not very large projects because of the absence of road pricing, and with very large projects, maintenance work would be too much. Basically, Funktionsbauvertrag is suitable for the medium-size projects In principle, other infrastructure projects other than road projects could be delivered under Funktionsbauvertrag. In this case, well-defined requirements for each type of infrastructure should be developed
1.3	*For PPP projects, there is feasibility test in Bavaria. Is there a similar document for Funktionsbauverträge?* There is no such separate feasibility test document for Funktionsbauvertrag. Federal Ministry of Transport, Building and Urban Development (BMVBS) developed such projects as an experiment, and these are implemented as pilot project. But there are some requirements and standard to be fulfilled to be implemented under Funktionsbauvertrag (e.g., the road section should be at least 10 km). Funktionsbauvertrag are till today implemented in the field of road superstructure
1.4	*What distinguishes the risk allocation in PPP projects and Funktionsbauverträgen?* Funktionsbauverträge are till now implemented in the field of superstructure, so the risks related to ground (Grundbaurisiko) are not transferred to the contractor which is case of PPP transferred to the contractor (private partner) because PPP includes construction of whole section of road. Traffic risks and traffic volume risk are not transferred to the contractor with Funktionsbauvertrag, but these risks are transferred to the contractor with PPP because with PPP as traffic volume increases, more money is collected through the road tolls that could be used for maintenance tasks. Funktionsbauvertrag being a fixed price contract such risks are not borne by the contractor
2	*Development and procurement*
2.1	*After first two pilot projects (A 81 in Baden-Württemberg: Oberndorf-Rottweil; and A 61 in Rhineland-Pfalz: Koblenz-Kruft) in 2002, only more three projects under Funktionsbauvertrag were implemented, and three more are on consideration until 2010; how do you see this development?* The development is relatively slow looking at the number of road projects in Germany. Functional requirements of Funktionsbauvertrag are similar to the requirements of PPP (Betriebsmodell), now the problem is to identify if the requirements of PPP are correct/appropriate for new type of contract "Funktionsbauvertrag." The development is also dependent upon the political decision. With PPP, the finance comes from the private sector, so PPP is more interesting for the politicians compared to Funktionsbauvertrag
2.2	*The awarding of PPP projects is usually based on a "competitive dialogue." Why was in Funktionsbauverträgen "restricted procedure" (nicht offenes Verfahren mit öffentlichem Teilnahmewettbewerb) (e.g., project A 61) used?*

(continued)

(continued)

	In this sense, Funktionsbauvertrag is more open for the public participation compared to PPP. Every construction company fulfilling the basic requirements could participate in bidding as with PPP. With PPP, among the bidders, the top four to six candidates are selected for the next round, and among them, the last two are selected for the final decision, but with Funktionsbauvertrag, all companies fulfilling the set requirements are selected for the next round selection, and in the end, the least price bidder is selected. The bid price is the sum of all three contract parts (A, B, and C) because these parts are not awarded separately
2.3	*Do you consider it meaningful to implement and award contract parts A, B, and C separately?* Funktionsbauvertrag is divided into parts A, B, and C, with their unique characteristics. Payment of part A is like conventional contract (based on unit price); part B includes functional requirements, and payment is on the installments, whereas payment with part C is done every 3 years. Parts B and C are strongly interconnected. Such properties make little sense to award and implement parts B and C separately. However, part A could be separately implemented and awarded from parts B and C
2.4	*How is the experience of Funktionsbauvertrag in Bavaria with project A 93, Section Brannenburg-Kiefersfelden under Funktionsbauvertrag?* The experience of A 93 under Funktionsbauvertrag is good. The construction has worked well without defects. The first inspection of the maintenance has been conducted which showed no defects. The inspection was conducted on the basic of visual inspection and the measurement techniques
3	*Finance, payment, penalty, and compensation*
3.1	*How is the payment with performance-based Funktionsbauvertrag calculated and paid?* With part A, the payment is made on completion on the basic of unit price. With part B, normally 90 % of the payment is made after the completion of construction, and the rest 10 % is paid after the handover inspection in the case if the project meets the defined requirements (however, these percentages could be different in different cases according to the agreement). With part C, the payment is normally made every 3 years after inspection and on meeting or exceeding the defined maintenance standard
3.2	*How is penalty criteria formulated? Are there clear rules of punishment?* There are of course penalty criteria with Funktionsbauvertrag. For noncompliance or not meeting standards, the contractor does not receive the payment. He should perform additional tasks to meet the defined standards. With part B, there is no penalty in terms of money, but the defined requirements should be met by the contractor during the handover inspection. In case of not meeting the standard, the contractor must perform the additional work to fulfill the requirements; otherwise, the rest of the payment is not made to the contractor. With part C, payment is done every 3 years after inspection. No defects should be seen during inspection; otherwise, the contractor is supposed to remove all the defects. With additional maintenance work, the contractor should maintain to fulfill the defined standard
3.3	*How do you assess the freedom of private partner regarding designing and construction for the possible cost optimization?* The private partner has freedom to choose the material requirements and techniques and use his experience for the implementation of construction cost optimization methods. But there is little room for the private sector to be innovative to reduce costs because innovation means risks and risks cost money. Functional requirements are already predefined, so at any cost, the defined standard has to be reached. Therefore, the contractor would follow the usual way rather than choosing new techniques and technologies because they could bring unnecessary risks

(continued)

(continued)

	If the contractor and the owner would work together sharing the arising risks, innovative approach would be encouraged
3.4	*How do you rate the financing by the private partner in PPP projects in contrast to the financing scheme in Funktionsbauverträgen?* PPP is financed through the money of the private sector. The public fund is limited, so PPP is very interesting for the politicians and the government. The interest rate for the private sector in PPP is higher, so there could be higher overall project costs compared to the government invested Funktionsbauvertrag projects. Funktionsbauvertrag do not suffer from the fluctuation of interest rate and the financial risk
3.5	*Is it possible to finance Funktionsbauvertrag through the private fund?* This case is very difficult. If finance is through the private fund, it is then no more Funktionsbauvertrag. In the case of private finance with Funktionsbauvertrag, financial risk arises and the fluctuation of interest rate plays vital role because it is a long-term contract. With the current concept of Funktionsbauvertrag, it is difficult to finance Funktionsbauvertrag projects through the private fund
3.6	*Has Funktionsbauvertrag achieved cost reduction?* Reasonable cost reduction has not been experienced. With construction, the cost reduction has not been noticed. But with maintenance, some costs have been reduced because of the reduction of personal and administration costs. But this amount is very small amount compared to the overall road project costs
4	*Comparison with performance-based road management and maintenance contract (PMMR)*
4.1	*There is a similar contract PMMR which is implemented in Australia, Canada, Sweden, Finland, the UK, etc., and is popular in South America. So why Germany didn't adopt it but develop Funktionsbauvertrag?* This contract type has been applied in Thüringen in state road (Landesstrasse) but with operation. In Germany, we want to develop such contracts ourselves rather than adopting from other countries (personal opinion)
4.2	*PMMR has been applied since 1988, and it has well-defined functional requirements and penalty criteria in terms of money (e.g., no potholes >2 cm deep on paved roads, penalty 110$/day/pothole). Does Funktionsbauvertrag have such criteria?* As said earlier on noncompliance under Funktionsbauvertrag, payment is not made, and the contractor should rework to meet the predefined standard. Such penalty in terms of money as with PMMR is not the case with Funktionsbauvertrag
4.3	*Germany is rather small country with well-developed road network. Germany needs more road maintenance and management programs rather than new road construction. Could the Funktionsbauvertrag concentrate on road management and maintenance as PMMR rather than on new road construction?* When we see the number of new road construction, road extension projects, or expansion of four-lane highway to six-lane highway, there is a need of project delivery methods for new road construction. Funktionsbauvertrag could be very useful in this context. Recently on highway A 6, Funktionsbauvertrag has been applied for the extension of the highway. There is also consideration of new road construction (extension of A 94) under Funktionsbauvertrag which is about 53 km
	End at 12:00
	Note: Interview is translated from German to English by Ashish Gajurel (the author of the book)

Appendix C. Interview Protocol with Mr. Stefan Högenauer and Dr. Bernd Zanker

Location of interview:	Oberste Baubehörde, München
Time:	01 July 2010, 10:00 to 12:00
Protocol taker:	Mr. Benno Vocke
Teilnehmer:	Mr. Stefan Högenauer, Oberste Baubehörde, München Dr. Bernd Zanker, Oberste Baubehörde, München Mr. Vocke, Research Associate, TU München Mr. Gajurel, Student TU München
Distributor:	All participants

Nr.	*Protocol points*
	Roads are the basic needs of mobility. The human interactions as well as economic activities are dependent on how well road infrastructures are developed. On the one hand, road construction is expensive, and on the other hand, it takes much time. But the governments of developed and developing countries have limited budget for road infrastructure development because the public fund should cover all the expenses and investments of the state. The public funds should cover sectors like health, education, agriculture, poverty reduction, power supply, and water supply. To overcome such financing and time constraint, various road project delivery methods (e.g., design-bid-build and design-build) are developed. With time, other delivery methods are evolved with eliminating the disadvantages of the previous delivery methods and with new process, procedures, and principles for effective and efficient project delivery. In Germany, Funktionsbauvertrag has been developed for the road project delivery. It has been developed with the aim to reduce the costs with quality improvements by expanding the role of private sector (design (plan), construction, and maintenance) and financing through the public fund
1	*Funktionsbauverträge in general*
1.1	*What are the reasons in your opinion for and against the choice of a Funktionsbauverträge in comparison to conventional procurement?*
	Funktionsbauvertrag includes maintenance work which is not included with conventional procurement. Funktionsbauvertrag is long term in nature compared to conventional
	What are the reasons in your opinion for and against the choice of a Funktionsbauverträge in comparison to the procurement under a PPP model?
	Funktionsbauvertrag has lower transaction costs compared to PPP. PPP projects are larger in nature compared to Funktionsbauvertrag. Therefore, PPP has more contracts with different banks, consultants, and subcontractors. PPP is financed by the private sector. The interest rate on private borrowing is higher than for public borrowing. Thus, PPP suffers from high interest rate of banks which may raise the overall project costs. Funktionsbauvertrag doesn't suffer such interest rate because it is financed through the public fund
	PPP is normally completed on time (Termintreue) because the contractor needs to collect his/her investment through the operation.
1.2	*Which project characteristics speak for or against the application of a Funktionsbauverträge?*
	There are no clear characteristics that speak for or against the application of Funktionsbauvertrag. But medium-size projects have been implemented under Funktionsbauvertrag

(continued)

(continued)

1.3	*For PPP projects, there is feasibility test in Bavaria. Is there a similar document for Funktionsbauverträge?*	
	There is not yet formalized feasibility test document for Funktionsbauvertrag. There is consideration to develop economic feasibility (wirtschaftliche Unterschung) test procedures with Funktionsbauvertrag. But some basic requirements are defined to consider project under Funktionsbauvertrag (e.g., length of road section).	
1.4	*How is Funktionsbauvertrag awarded? What are the selection criteria?*	
	Funktionsbauvertrag has been awarded through restricted procedures ("nicht offenes Verfahren mit öffentlichem Teilnahmewettbewerb"), whereas selection with PPP is through open dialogue. With Funktionsbauvertrag, every interested company fulfilling basic requirements can send their application depending upon the published offer. In the second phase, the owner invites limited company for bidding. The owner normally evaluates the application depending upon the experience, past performances, and achievements	
3	*Penalty and nature*	
3.1	*How is penalty criteria formulated? Are there clear rules of punishment?*	
	There penalty criteria with Funktionsbauvertrag. For noncompliance or not meeting standards (contract part A and part B), the contractor payment is deducted, or he should perform additional tasks to meet the defined standards. Penalties in terms of money are not applied	
3.2	*How do you assess the freedom of private partner regarding designing and construction for the possible cost optimization?*	
	Innovation is limited with Funktionsbauvertrag because it has been applied only in the area of road superstructure. The area of work is limited to superstructure. How thick the superstructure should be and what should be its color, its slope, evenness, etc., are already defined. So with Funktionsbauvertrag, the contractor has limited innovative possibility PPP for road project includes earthworks, inner layer, outer layer, and all other aspects of road. So with PPP, there are lots of areas where innovative approach could be implied. But PPP loans its large part of investment from the bank. So bank wants that the company bears less risk. Using innovation means accepting all arising risks. Because of this condition, PPP also cannot use innovative approach. PPP can, for example, use innovative approach with bridge construction where it can optimize the costs in selection of material for support pillar. In general, Funktionsbauvertrag and PPP have less effect on designing and in implementation of innovative approach	
3.3	*The structure of PPP is combined with special purpose vehicles (Projektgesellschaft). Is it possible to combine Funktionsbauvertrag structure with special purpose vehicles?*	
	Funktionsbauvertrag doesn't have all elements of PPP. Funktionsbauvertrag has less number of contracts and partners compared to PPP. With PPP, the owner doesn't want to contact with private partners, banks, and other consultants. In such case, SPV is important as a single contract partner for the public sector. Till now, structure of Funktionsbauvertrag does not have SPV. In case of private finance, SPV is important. In the future, however, research could be conducted if SPV would be useful for Funktionsbauvertrag	
	End at 12:00	
	Note: The interview protocol is translated from German to English by Ashish Gajurel (the author)	

Appendix D. Funktionsbauvertrag (A 93: Brannenburg-Kiefersfelden)[1]

Introduction

It is the third road project in Germany under Funktionsbauvertrag. The initial work of the project started in 2003, and the construction work started in 2004. The project is the section on the federal highway A 93 in Bayern.

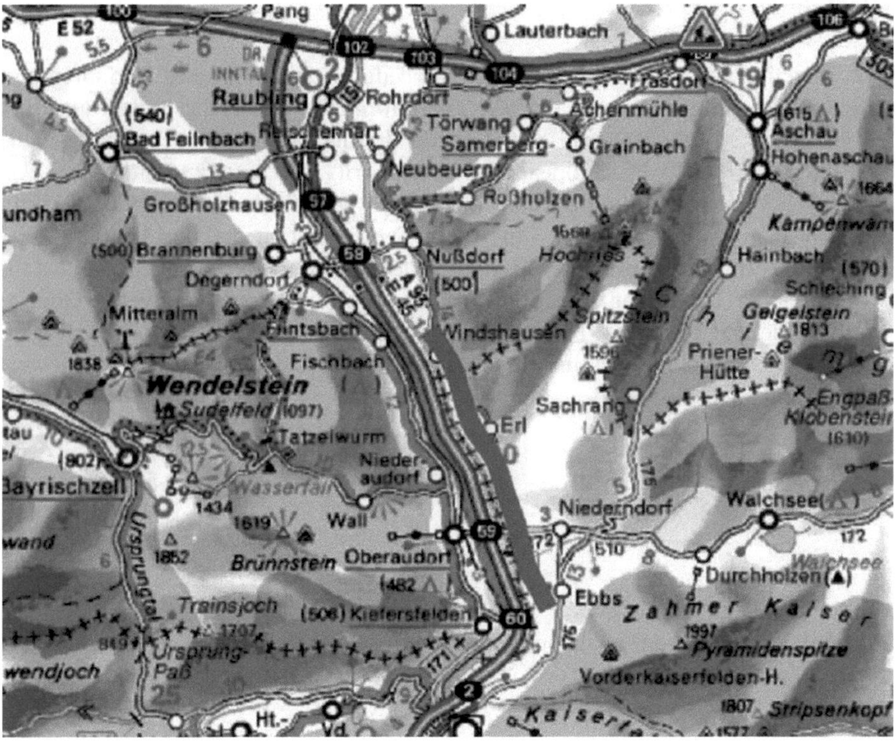

Fig. A.D.1 Site map A 93 Süd Rosenheim-Kiefersfelden (Schmerbeck, Rupert: Pilotprojekte mit Funktionsbauvertragen – Vertragliche Sicherung der Oberflächeneigenschaften, München, p. 69, http://www.vsvi-hessen.de/download/20051102/vsvi2005_11_02schmerbeck.pdf. Accessed 26 Mar 2010.)

[1] The whole chapter is based on the Schmerbeck, Rupert: Pilotprojekte mit Funktionsbauvertragen – Vertragliche Sicherung der Oberflächeneigenschaften, München, http://www.vsvi-hessen.de/download/20051102/vsvi2005_11_02schmerbeck.pdf. Accessed 26 Mar 2010 and Schmerbeck and Löcherer: Vermerk Bundesautobahn A 93 Süd Rosenheim – Kiefersfelden AS Brannenburg – AS Kiefersfelden, Vermerk über die Erfahrungen bei der Vergabe und Bau, Autobahndirektion Südbayern, June 2008, München.

The contract comprises the renewal of the 11.7 km superstructure. The renewal work was with the concrete floor with exposed aggregate surface, leading to a noise reduction in a similar magnitude. The quality of the road is defined on the perspective of the road users. The contract included the maintenance with renewal of the existing road section. The contract period is 20 years. The Autobahndirektion Südbayern reported about 12 km long section A 93 Süd to the higher construction authority (Oberste Baubehörde) to be renewed. In 2002, it was decided to renew this section under Funktionsbauvertrag.

Selection

The selection was based on restricted procedure ("nicht offenes Verfahren mit öffentlichem Teilnahmewettbewerb"). The information about the project was first published on 30 July 2003 in Federal Tender Journal (Bundesausschreibungsblatt) and on 30 July 2003 in European Union Authority Journal (EG-Amtsblatt). The requirements to be fulfilled by the bidder were published on 21 August 2003 in EG-Amtsblatt, on 15 August 2003 in Federal Tender Journal, and on 14 August 2003 in State Information Journal. The bidding was open to the potential candidates till 17 September 2003. The selection decision was made together by the Oberste Baubehörde and BMVBM. Fourteen construction companies bade for the project. The selected candidate was informed on 13 January 2004. In the end, the lowest price bidder was selected.

Contract Parts

Part A

Part A works as conventional contract where the payment is based on the unit price.

Part B

With the past Funktionsbauverträge in part B, the reference for construction standard is provided by the owner to the contractor. The contractor is required to construct the road meeting or exceeding the provided requirements. Part B included the construction of superstructure.

Part C

Part C includes the maintenance work. The payment is not paid every year but every 3 years on annuity basic. A 2 % interest rate is considered on the payment.

Experience of the Project

Performance

The quality of the constructed and executed superstructure is evidently very good. The constructor implemented their experience and qualified personnel. The superstructure is very even maybe because of the concrete floor with exposed aggregate surface (Waschbetonoberflächenstruktur). The overhand inspection was done, but the inspection on the grip quality of total section under contract couldn't be completed because of the change of the season. The inspection values showed the positive results of the condition of the superstructure. The grip value of the measured section is excellent and showed the middle values of approximately 0.6 μ_{SCHRIM}.

Effort in the Field of Construction Supervision

According to the field construction manager, the road is in good condition that it doesn't need the supervision. Some small inspection was done in the area of superstructure. In the real sense, in the area of superstructure, the control test according to the ZTV Concrete was not executed, although the protocol of the inspection was demanded from the contractor.

Contractual Settlement (Vertragliche Abwicklung)

There is no problem with part B in the context of contractual settlement as long as the functional requirements and the precise design of the superstructure are well defined. The payment of part B is based on installments. With this project, about 50 % of the payment was made on the completion of construction and the rest after the handover inspection. But this process has very little sense to delay the payment because of not carried inspection of the small section. It would make good sense to make payment of about 70 % after the completion of construction and inspection of visual defects and the remaining 30 % after the successful results of inspection.

Experience of the Maintenance Period

Till now no defect has been detected. In 2007, the first functional inspection was carried. The inspection report of contractor showed no defects. The inspection commission showed that the recognizable defects were not seen. The result of the inspection was submitted to the owner which presented no defects in sense of ZTV Funktion which defines the performance standard.

Glossary

Project delivery systems It is a process through which a construction project is comprehensively designed and constructed for an owner. It includes the project scope definition; organization of designers, constructors, and consultants; sequencing of design and construction tasks; execution of design and construction; and closeout and start-up.[2]

Design-bid-build It is a project delivery method where the design team completes the design first; then the owner contracts with the general contractor to build the project depending on that design.[3]

Design-build It is a project delivery method where the owner contracts with the single entity to design and construct the project; the contractor could, however, be an architect, developer, or other entity.[4]

Construction manager at risk It is a project delivery method where the contractor is usually referred as the construction manager. The construction manager assists the owner and the design team and is involved in design and planning phase as well as responsible for the construction of the project.[5]

Performance-based contract It is a contracting method where the owner does not specify any method or material requirements as long as country's standards are met; instead, he specifies performance level that the contractor should meet when delivering the project (output) or services.[6]

Key performance indicator (KPI) KPIs are a set of key control parameters which give a measure of the performance of the project or services.[7]

[2] Loulakis MC. Design build for public sector. New York: Aspen Publishers, Inc.; 2003. p. 106.

[3] Glavinish TE. Contractor´s guide to green building construction. Hoboken: Wiley; 2008. p. 36.

[4] Glavinish TE. Contractor´s guide to green building construction. Hoboken: Wiley; 2008. p. 37.

[5] Glavinish TE. Contractor´s guide to green building construction. Hoboken: Wiley; 2008. p. 36.

[6] Stankevich N, Qureshi N, Queiroz C. Performance-based contracting for preservation and improvement of road assets. Washington, DC: The World Bank; September 2005 (updated August 2009). p. 1.

[7] Turner RJ. The handbook of project based management – leading strategic change in organizations. New York: McGraw-Hill; 2009. p. 52.

A. Gajurel, *Performance-Based Contracts for Road Projects: Comparative Analysis of Different Types*, DOI 10.1007/978-81-322-1302-4, © Springer India 2014

Service-level agreement (SLA) SLA is a written agreement between the owner and the contractor which includes the definition of service level, performance standard, problem management, owner duties and responsibilities, warranties, legal compliance, and the condition of the termination of agreement.[8]

Performance-based contracts for the maintenance and management of roads (PMMR) A PMMR is an agreement between an owner and a private contractor, whereby the private contractor maintains and manages the road to achieve specified condition standards for a certain period of time for a fixed payment.[9]

Funktionsbauvertrag (FBV) FBV is a construction contract between an owner and a private contractor for the new road construction or renovation, management, and maintenance of the road, where the quality of work is no longer described by the materials and construction techniques rather over the function from the prospective of road user defined.[10]

PPP PPP is a form of cooperation between public and private sectors for the funding, construction, renovation, management, or maintenance of an infrastructure or services.[11]

Special purpose vehicles (SPV) PPP project involves a public sector agency and a private sector consortium with contractors, maintenance companies, private investors, as well as consulting firms. The consortium often forms SPV. The SPV signs contracts with the government and subcontractors to build and maintain the facility.[12]

Public sector comparator (PSC) The PSC is an evaluation as well as a benchmarking tool used in the assessment of the project (could be PPP project) to decide which model is cost-efficient.[13]

Risks Risk is the uncertainty of an assumption relating to its appropriateness to achieve future objectives and contains the potential of chances as well as dangers depending on the certainty of this assumption.[14]

Process and procedures A procedure is a complete operation; that means it is the complete set of action that results in some desired operation, whereas a process

[8] Blokdijk G. Agreement 100 success secrets, SLA, Service Level Agreements. Service Level Management and Much More; 2008. p. 175.

[9] www.adb.org/Documents/Supplementary-Appendixes/39676/Supplementary-Appendix-A.pdf. Accessed 24 Apr 2010.

[10] Schmerbeck R. Pilotprojekte mit Funktionsbauvertragen – Vertragliche Sicherung der Oberfläche-neigenschaften, München, p. 70. http://www.vsvi-hessen.de/download/20051102/vsvi2005_11_02schmer beck.pdf. Accessed 26 Mar 2010.

[11] Zimmermann J. Project delivery systems, Lecture note in Lehrstuhl für Bauprozessmanagement und Immobilienentwicklung an der Technischen Universität München, Issue Jan 2009. pp. 7–28.

[12] ADB. Facilitating Public-Private Partnership for accelerated infrastructure development in India. Workshop report, December 2006, p. 16 and 17.

[13] Zimmermann J. Project delivery systems, Lecture note in Lehrstuhl für Bauprozessmanagement und Immobilienentwicklung an der Technischen Universität München, Issue Jan 2009. p. 9 and 10.

[14] Zimmermann J. Project delivery systems, Lecture note in Lehrstuhl für Bauprozessmanagement und Immobilienentwicklung an der Technischen Universität München, Issue Jan 2009. p. 3.

is the series of individual steps within the procedures that is followed to achieve the objective.[15]

Effect and affect Effect is mostly used as a noun and refers to something that happens due to some action or event. Its good synonym is "result," whereas affect is mostly used as a verb and refers to the action of influencing something else. Its good synonym is "influence."[16]

Efficient and efficiency Efficient is being able to know what is the right thing to do and actually doing it or determination of the kind of tool that is right for the mission is being efficient, while efficiency is being able to use the resources or tools available successfully to achieve the goal.[17]

International Rough Index (IRI) IRI is an index for the description of roughness or smoothness of the road pavements. It was developed by the World Bank in 1982. It is important to be known because pavement smoothness affects the dynamics of moving vehicles, impacting the rate of deterioration of the pavement and the operation and safety of vehicle, river, and passenger.[18] It is calculated as the total anticipated vertical movement of the vehicle accumulated over the length of the section; the typical unit of IRI is meters/kilometer or inches/mile. If pavement is perfectly smooth, IRI would be zero.[19]

Rehabilitation, renovation, and maintenance[20] Rehabilitation is to restore to the former condition.

Renovation is to restore to a good state of repair (make new again).

Maintenance is to keep in a good condition by checking or repairing it regularly.

---

[15] Adar BS. Technical writing – the difference between a process and a procedures. http://ezinearticles.com/?Technical-Writing---The-Difference-Between-a-Process-and-a-Procedure&id=1261798. Accessed 13 May 2010.

[16] Bowman D. http://ezinearticles.com/?Effect-Vs-Affect&id=973048. Accessed 13 May 2010.

[17] Moon J. http://ezinearticles.com/?Effectiveness-Vs-Efficiency&id=3963119. Accessed 13 May 2010.

[18] Jackson NM. Preliminary report – an evaluation of the relationship between fuel consumption and pavement smoothness, University of North Florida, USA, 2004, p. 1.

[19] Janisch D. An overview of Mn/DOT's pavement condition rating procedures and indices, Gervais Avenue, March 2003, p. 3.

[20] Meaning from the Concise Oxford English Dictionary, Eleventh Edition, Revised.

Bibliography

ADB. Asian Development Bank: Public-Private Partnership handbook. Philippines.

ADB. Face-lifting Public-Private Partnership for accelerated infrastructure development in India. Workshop report, December 2006.

ADB. Performance-based maintenance by contracts, Supplementary appendix A, P. 1.

Akintoye A, Beck M, Hardcastle C. Public-Private Partnership: managing risks and opportunities. New York: Wiley and Sons Inc Publication; 2003.

Alfen Consult GmbH Weimer; PPP- Businessmodelle im Überblick, Weimar, Germany, March 2008.

Alfen HW, Leupold A. Public Private Partnership (PPP) in the German Public Real Estate Sector, Germany real estate – yearbook 2007. The Hague: Europe Real Estate Publisher B.V; 2007.

Allen G. The Private Finance Initiative (PFI). Research paper 01/117, UK, December 2001.

Altmüller P. Schriftenreihe Bauwirtschaft – Tagungen und Berichte 6. Kassel: Kassel University Press; April 2009.

American Water Works Association. Construction contract administration, Denver, USA; 1999.

ASCE. Alternative project delivery, procurement, and contracting methods for highways. USA; 2007.

Atkin B, Brooke A. Total facilities management. Oxford: Blackwell Science; 2000.

American Water Works Association (AWWA) Manual. Construction contract administration. Denver: AWWA Publication; 1996.

Ballati D. Privatizing Governmental functions. New York: Law Journal Press; 2004.

Bangemann TO. Shared services in finance and accounting. Aldershot: Gower; 2005.

Baroudy E, Abid AL, Attia B. Managing water demand – policies, practices and lessons from the Middle east and North Africa Forums. London: IWA Publishing; 2005.

Beckers T. Die Realisierung von Projekten nach dem PPP-Ansatz bei Bundesfernstraßen, Dissertion, TU Berlin, Fakultät Wurtschaft & Management, Berlin; 2005.

Beckers T, Klatt JP. Zeitliche Homogenisierung und Berücksichtigung von Risiko im Rahmen von Wirtschaftlichkeitsuntersuchungen, Berlin, November 2009.

Beckers T, Hirschhausen CV, Klatt JP. Reformbedarf bei den Bundesstrassen und das potential des PPP-Ansatzes, Berlin, May 2009.

Bekanntmachung der Neufassung der Vergabe- und Vertragsordnung für Bauleistung (VOB) Teile A und B, September 2002.

Beyond the 1990s. Ensuring quality in an era of limited resources – summary of symposium-Federal Facility Council, standing committee on Organization and Administration, USA; 1997.

Biller N. Presentation of the Münchner Massivbau-Kolloquium Sommersemester 2009, München.

Blaiklock TM. Public – Private Partnership ("PPP") – the advantages and limitations, Berne, January 2005.

Blokdijk G. Agreement 100 success secrets, SLA, Service Level Agreements. Service Level Management and Much More; 2008.

A. Gajurel, *Performance-Based Contracts for Road Projects: Comparative Analysis of Different Types*, DOI 10.1007/978-81-322-1302-4, © Springer India 2014

Brian C, Fekpe E, Gopalakrishna D. An evaluation of the use of innovative performance specification in highway construction. Final report, Washington, DC, February 2003.
Brown PC. Implementing SOA – total architecture in practice. Boston: Boston Harvard Business School Press; 2008.
Burki SJ, Perry G, Dillinger W. Beyond the center: decentralizing the state. World Bank Latin American and Caribbean Studies. Washington, DC: World Bank; 1999.
Burningham S, Stankevich N. Why road maintenance is important and how to get it done. The World Bank Document, TRN-4, June 2005.
Byoungki KIM. Infrastructure development for the economic development in developing countries, lessons from Korea and Japan. Kobe: Kobe University; November 2006.
Cabana G, Liautaud G, Faiz A. Areawide performance-based rehabilitation and maintenance contracts for low – volume roads. Transportation research record 1625. New Delhi: Central Road Research Institute.
Camarinha-Matos LM, Paraskakis I. Leveraging knowledge for innovation in collaborative networks, Greece, October 2009.
Cameron Jr JG. A practitioner's guide to construction law. Philadelphia: ALI-ABA Publisher; 2000.
Charlier F, Cho-Oguie C. Sustainable reforms for inclusive growth in Cameroon – a development policy review. Washington, DC: The World Bank; 2009.
Cibinic Jr J. Performance-based contracting. Washington: Department of Energy; 2000.
Cleland, DI, Ireland LR. Project manager's portable handbook, USA; 2004 Contract information Bulletin.
Click RL, Duening TN. Business process outsourcing, the competitive advantage. Hoboken: John Wiley and Sons Inc Publisher; 2005.
Competitive dialogue; the EU new procurement procedures. London: Freshfields Bruckhaus Deringer; 2005.
Competitive dialogue procedure. London: Office of Government Commerce Publication; 2006.
Construction Management JumpStart, San Francisco; 2004.
Construction Specifications Institute. The project resource manual – CSI manual of practice. 5th ed. New York: McGraw-Hill; 2005.
Cunningham Greg AIA. Commissioning large public project using construction manager at risk (CM@R). National conference on building commissioning 4–6 May 2005.
Cushman RF, Loulakis M. Design-build contracting handbook. New York: Wiley; 2001.
Cushman RF, Cushman KM, Cook SB. Construction litigation: presenting the owner. New York: Aspen Publishers; 1990.
Deloitte Research Study 2006. Closing the infrastructure gap. Role of Public-Private Partnerships. New York: Deloitte; 2006.
Demkin JA (Bernstein PG, FAIA, RIBA, LEED AP), The American Institute of Architects. The architect's handbook of professional practice. New Jersey: Wiley; March 2008.
Deutsches Baublatt. Issue November 2008, 35. Jahrgang, Nr. 340.
Dreher A. Pilotprojekte mit Funktionsbauverträgen Erwartungen der Strassenbauverwaltung, Strasse + Autobahn 5, NO. 5, Germany; 2003.
Dreher A. Pilotprojkte mit Funktionsbauvertraege – Erwartung der Strassenbauverwaltung, Koblenz, Mai 2003 P. 260 (Strasse+Autobahn, 54 Jahrgang).
Eggers WD, Startup T, Deloitte Research. Closing the infrastructure gap: the role of Public-Private Partnerships. New York: Deloitte; 2006.
Eifert H. Funktionsbauverträge – ein Modell der Zukunft? Köln, April 2004.
Estache A, Carbajo JC, de Rus G. Argentina's transport privatization and re-regulation: ups and downs of a decade long daring experience. 2nd ed. Washington, DC: The World Bank; March 1999.
Ewers K. Assessing risk for health care IT organizations – a practical overview approach, February 2007.
Executive Summary. Design-build effectiveness study. Final report. US Department of Transportation, Federal Highway Administration, January 2006.
Forschungsgesellschaft für Strassen – und Verkehrswesen, Arbeitsgruppe Fahrzeug und Fahrbahn: ZTV Function-StB 01, Koeln; 2001.

Forsthoffer W. Reliability optimization through component condition monitoring and root cause analysis. Oxford: Elsevier; 2005.
Friedlander MC, Roberts KM. Design-build – the single entity option. Chicago: SCHIFF HARDIN LLP; January/February 1997.
Geddes M. Making Public Private Partnerships works, building partnerships and understanding cultures. Burlington: Gower; 2005.
Gerdes E. PPP-Ansätze für Fernstrasse. Dissertation an Universität Hamburg, December 2007.
Geschäftsbericht 2007. Ministerium für Bauen und Verkehr des Landes Nordrhein-Westfalen, Strassen.NRW, p. 30.
Glavinich TE. Contractor's guide to green building construction: management, project delivery, documentation, and risk reduction. Hoboken: Wiley Printing; 2008.
Grimsey D, Lewis MK. Evaluating the risks of Public Private Partnerships for infrastructure projects, Australia, July 2000.
Grimsey D, Lewis MK. Public Private Partnership, the worldwide revolution in infrastructure provision and project finance. Northampton: Edward Elgar; 2004.
Güngerisch A. Neue Bau- und Beschaffungsverträge. USIC seminar, 15 October 2008.
Hans WA, Andreas L. Public Private Partnership (PPP) in the German Public Real Estate Sector, Yearbook 2007.
Heng TM, Yam TK. Competitiveness of the Singapore economy, a strategic perspective. Singapore: Singapore University Press; 1998.
Higgins P. Institution of Highways and Transportation. Relationship contracting – contracts and procurement designed to achieve project objectives, UK, May 2009.
Hodge G, Greve C. Public-Private Partnership, learning from international experience. Cheltenham: Edward Elgar; 2005.
IMF. Public Private Partnerships, government guarantees, and financial risks. Washington, DC: IMF; 2006.
Institute of Chemical Engineers. Form of contracts, lump sum contracts. 4th ed. Rugby: Institute of Chemical Engineers; 2001.
Jackson BJ. Construction management JumpStart. 2nd ed. San Francisco: SYBEX Inc. Publication; 2004a.
Jackson NM. Preliminary report – an evaluation of the relationship between fuel consumption and pavement smoothness. Jacksonville: University of North Florida; 2004b.
Janisch D. An overview of Mn/DOT's pavement condition rating procedures and indices. Gervais Avenue, March 2003.
Jansen SA, Priddat BP, Stehr N. Die Zukunft des Öffentlichen. Wiesbaden, January 2007.
Jordan A. Presentation of managing risks in PPP projects through legal documentation, Jordan, September 2007.
Kappel F. Pilotprojkte mit Funktionsbauverträgen – Erwartungen der Bauwirtschaft, Strasse + Autobahn, Germany; 2003.
Kerzner H. Project management – a system approach to planning, scheduling and controlling. Hoboken: Wiley; 2009.
Khan NA. Infrastructure for economic development: a comparative study of India and Malaysia. New Delhi: Anmol Publications Pvt; 2004.
Kwak ZH, Randall B. Construction management at risk: an innovative project delivery method at stormwater treatment area in the Everglades, Florida, p. 2.
Kymell W. Building information modelling, planning and managing construction projects with 4D CAD and simulation. New York: McGraw-Hill; 2008.
Lang A. Investor – Hochschule – Bauindustrie, Kolloquium 2006, Muenchen, under the topic PPP.
Levy SM. Construction superintendent's operations manual. New York: McGraw-Hill; 2004.
Levy SM. Design build project delivery. New York: McGraw-Hill; 2006.
Liautaud G. Maintaining roads – experience with output-based contracts in Argentina. The World Bank Report 2004.
Local Agencies and Design-Build Contracting: A Briefing paper for Legisltors and their Staff, California; 2009.

Loulakis MC. Design build for public sector. New York: Aspen Publishers, Inc.; 2003.
Manual of water supply practice – M47: construction contract administration. USA: American Water Work Association; 1997.
Matthew Bender (1992), Cameron JG Jr. A practitioner's guide to construction law, USA; 2000.
Mincks WR, Hal J. Construction jobsite management. 2nd ed. New York: Thomson Delmar Learning Publication; 2004.
Moelmann LR, Harris JT. The law of performance bonds. Chicago: American Bar Association Publication; 1999.
Molenaar KR, Yakowenko G. Alternative project delivery, procurement, and contracting methods for highways. Reston: ASCE; 2007.
National Cooperation Highway Research Program. Guidebook to highway contracting for innovation: the role of procurement and contracting approaches in facilitating the implementation of research findings, USA, 1999.
NCHRP Report 428. Guidebook to highway contracting for innovation: the role of procurement and contracting approaches in facilitating the implementation of research findings, Washington, DC.
NCHRP Synbook 389. Performance-based contracting for maintenance. Washington, DC; 2009.
NCHRP Synbook 391. Public sector decision making for public-private partnerships, 2009 Transport Research Board NCHRP Synbook 389: Performance-based contracting for maintenance, USA, 2009.
Nösler I. Oberbaubemessung im Rahmen neuer Bauvertragsformen (Funktionsbauvertrag). Brüssel, p. 2.
Nösler I. Pavement Design in Rahmen neuer Bauvertragsformen. Brüssel, tis October 2003.
OECD. Performance indicators for road sectors: summary of the field test. Paris: OECD; 2001.
OECD. Public-Private Partnerships: in pursuit of risks sharing and value of money. Paris: OECD; 2008a.
OECD. Transport infrastructure investment: options for efficiency. Paris: OECD; 2008b.
Ohio Department of Transportation. Project development process manual. Arizona: Ohio Department of Transportation; 2004.
Ohrn GL, Roger TPE. Defining project Delivery Methods for Design, Construction, and other Construction – Related Services in the United States. Alabama: Auburn University; 2008.
Ozdemir L. North American tunneling. The Netherlands; 2002.
Pakkala P. Performance-based contracts – international experiences, Finnish Road Administration. Presentation at the TRB workshop on performance-based Contracting, Washington, DC, April 2005.
Parmenter D. Key performance indicators, developing, implementing, and using winning KPIs. Hoboken: Wiley; 2007.
Pauli C. Schriftenreihe Bauwirtschaft – Tagungen und Berichte, IBW-Symposium, Innovative Abwicklungsformen für Bauprojekte: Partnering with PPP, Kassel, 2006.
Performance – based contracting: here to stay, but challenges ahead, Fall 2004.
Performance based contracting: a viable contract option? Summary of conference, U.S. and international approaches to performance measurement for transportation systems. Conference proceedings 44, Washington DC.
Performance based contracts: global transport knowledge partnership. www.gtkp.com/theme.php?themepgid=118. Accessed 21 Mar 2010.
Performance-based management and maintenance of roads (PRRM) Case study 1: Argentina based on Toolkit PPP in highways, The World Bank.
Potts KF. Construction cost management; learning from case studies. London: Taylor and Francis Publication; 2008.
Pratt D. Fundamentals of construction estimating. 2nd ed. New York: Thomson Delmar Learning Publication; 2004.
Price ADF. Financing international projects. Geneva: International Labour Office; 1995.
Process in improving Project Management at the Department of Energy, National Research Council (US), Committee for Oversight and Assessment of U.S Department, Washington, DC.

Procurement of performance-based management and maintenance of roads. Washington DC: The World Bank; February 2002.

Procurement of performance-based management and maintenance of roads (Output-based service contract). The World Bank Report, Revised March 2003.

Public Private Partnership (PPP) guideline: Public-Private Partnership Unit, Prime Minister Department, Putrajaya, Malaysia.

Public Private Partnerships: An introduction. Research paper No. 1 2002–2003. Department of the Parliamentary Library, Commonwealth of Australia, 2002.

Quality in the constructed project: a guide for owners. Designers and Constructors USA (ASCE Manual No.73), 2000.

Racky P. PPP in Hessen – auch im Kommunalen Strassenbau? Summary of the presentation, Hessen, March 2003.

Racky P. Der Funktionsbauvertrag aus betriebswirtschaftlicher Sicht, VSVI – Seminar "Funktionsbauverträge", Friedelberg, April 2010.

Robinson H, Carrilo P, Anumba C, Patel M. Governance and knowledge-management for Public-Private Partnership. Malden: Wiley-Blackwell; 2010.

Rodrigue J-P, Comtois C, Slack B. The geography of transport systems. New York: Routledge Publication; 2009.

Sanvido VE, Kanchar MD. Project delivery systems, CM at risk, design-build, design-bid-build. Austin: The Construction Industry Institute – The University of Texas at Austin; 1998.

Satterfield Z. Tech brief. Publisher by the National Environmental Service Center, West Virginia University, Vol. 9, Issue 2; Summer 2009.

Schaefer S. Der Funktionsbauvertrag, Presentation 11, Lehrstuhl Bauproyessmanagement und Immobilenentwicklung Technische Universitaet Muenchen.

Schmerbeck R. Pilotprojekte mit Funktionsbauvertragen – Vertragliche Sicherung der Oberflächeneigenschaften, München, p. 70.

Schmidt H. Pilotprojkte mit Funktionsbauverträgen – Erste Praxiserfahrungen, FGSV, Schriftenreihe der Arbeitsgruppe, Bonn; 2004.

Scott S III. Performance – based contracting: a viable contract option? Summary of theconference, U.S. and international approaches to performance measurement for transportation systems, USA

Shah A. Public service delivery. Washington, DC: The World Bank; 2005.

Shiffmann F. IVT ETH Zurich, Performance-based specification for the operational and structural road maintenance of communal road network. Conference paper STRC 2007, Monte Venta/Ascona, September 2007.

Shrestha HR. Sustainable approach to road transport maintenance practice for effective service delivery. Policy paper 24, Kathmandu, September 2006.

Sidney S III, Konrath L. Performance based contracting: a viable contract option? Summary of conference, U.S. and international approaches to performance measurement for transportation systems. Conference proceedings 44, Washington DC.

Stanford University. Project management, A Department of Land, Building and Real Estate. Project delivery manual: Vols. 1 and 2.

Stankevich N, Qureshi N, Queiroz C. Performance-based contracting for preservation and improvement of road assets. Washington, DC: The World Bank; September 2005 (updated August 2009).

Strassen NRW Presseinformation.

Stuart C. Applied transport economics – policy, management and decision making. 3rd ed. London: Kogan Page Publisher; 2005.

Terry GR, Franklin SG. Principles of management. Homewood: R.D. Irwin; 1982.

Testa MF, Poertner J. Fostering accountability using evidence to guide and improve child welfare policy. New York: Oxford University Press; 2010.

The American Institute of Architects. The architect's handbook of professional practice. Hoboken: Wiley; March 2008.

The American Institute of Architects. Architectural graphic standards for residential construction. Hoboken: Wiley; 2010.

The National Research Council, Washington, DC.

Treasury HM. The green book. London: Appraisal and Evaluation in Central Government; 2003.
Tucker, C. Performance measurement: a technical prospective. Washington, DC, p. 1. http://unstats.un.org/unsd/dnss/QAF_comments/USA%20Bur%20of%20Labor%20Performance%20Measurement_Clyde%20Tucker%20Performance%20Measurement.pdf. Accessed 25 Apr 2010.
Turner RJ. The handbook of project based management – leading strategic change in organizations. New York: McGraw-Hill; 2009.
Universitaet Stuttgart, Kolloquium- Technische Absicherung des Regelwerks, Privatfinanzierter Strassenbau – Chancen und Risiken, Stuttgart, 2005.
Uher TE, Davenport P. Fundamentals of building contract management. Sydney: University of New South Wales Press Publication; 2009.
Usher TE, Davenport P. Fundamental of building contract management. Sydney: UNSW Press; 2002.
Verdi. PPP in Strassenbereich, April 2007.
Wearne S. Engineering management, civil engineering contracts. London: Thomas Telford Ltd Publishers; 1989.
Wiederspahn M. Brückenbau: 10. Symposium Brückenbau in Leipzig, February 2010.
Wolper LF. Health care Administration Planning, Implementing, and Manageing Organized Delivery Systems. 4th ed. Sudbury: Jones and Barlett Publishers; 2004.
World Bank. Performance based contracting for preservation and improvement of road assets. Washington, DC: World Bank; April 2006.
Yescombe ER. Public Private Partnerships, principles of policy and finance. Oxford: Butterworth-Heinemann; 2007.
Zietlow G. Cutting costs and improving quality through performance-based road management and maintenance contracts – Latin America and OECD experiences-, Birmingham, April 2004a.
Zietlow G. Implementing performance-based road management and maintenance contracts in developing countries – an instrument of German Technical Cooperation, Eschborn, November 2004b.
Zietsman J. Performance measure for performance based maintenance contracts. Research of Texas Department of Transportation, The World Bank document, p. 9.
Zimmermann J. Project delivery systems. Lecture note in Lehrstuhl für Bauprozessmanagement und Immobilienentwicklung an der Technischen Universität München, Issue Jan 2009.
Zimmermann J. PPP als ein Modell zur Übertragung von Steuerungsprozessen – Möglichkeiten bei Hochbau- und Infrastrukturprojekten, Munich, March 2010.

Websites

Adar BS. Technical writing – the difference between a process and a procedures. http://ezinearticles.com/?Technical-Writing---The-Difference-Between-a-Process-and-a-Procedure&id=1261798. Accessed 13 May 2010.
ADB. Performance–based maintenance by contracts. Supplementary Appendix A, p.1. www.adb.org/Documents/Supplementary-Appendixes/39676/Supplementary-Appendix-A.pdf. Accessed 24 Apr 2010.
Alfen HW, Leupold A. Offering new and integrating opportunities for the private sector. Public Private Partnerships (PPP) in the German Public Real Estate Sector, www.europe-re.com/system/main.php?pageid=2242&articleid=9378. Accessed 26 Mar 2010.
Bayerische Staatregierung. www.bavaria.de/Pressemitteilungen-.1255.10242857/index.htm. Accessed 31 May 2010.
Bowman D. http://ezinearticles.com/?Effect-Vs-Affect&id=973048. Accessed 13 May 2010.
Contract information bulletin. www.usaid.gov/business/business_opportunities/cib/pdf/cib91_18.pdf. Accessed 6 Apr 2010.

Executive summary, design-build effectiveness study. Final report. US Department of Transportation, Federal Highway Administration, January 2006. www.fhwa.dot.gov/reports/designbuild/designbuild.htm. Accessed 20 Mar 2010.

Kwak ZH, Randall B. Construction management at risk: an innovative project delivery method at stormwater treatment area in the Everglades, Florida, p. 2. http://home.gwu.edu/~kwak/asce_congress_6.pdf. Accessed 20 Mar 2010.

Ministry of Transport and Communication of Kyrgyz Republic. http://sk-road.com/performance_based_maintenance_pbm. Accessed 7 June 2010.

Moon J. http://ezinearticles.com/?Effectiveness-Vs-Efficiency&id=3963119. Accessed 13 May 2010.

Newcastle City Council, Australia. www.newcastle.nsw.gov.au/__data/assets/pdf_file/0009/92628/140-A.pdf, p.1. Accessed 11 Apr 2010.

Nösler I. Oberbaubemessung im Rahmen neuer Bauvertragsformen (Funktionsbauvertrag), Brüssel, p. 2. www.nynas.com/upload/docs/Bitumen/NOAH.pdf. Accessed 26 Mar 2010.

Ohio Department of Transportation. Project development process. www.nynas.com/upload/docs/Bitumen/NOAH.pdf. Accessed 26 Mar 2010.

Online Compact Oxford English Dictionary. http://www.askoxford.com/concise_oed/infrastructure?view=uk. Accessed 14 Mar 2010.

Performance based contracts: global transport knowledge partnership. www.gtkp.com/theme.php?themepgid=118. Accessed 21 Mar 2010.

Performance measurement: a technical prospective, Washington, DC, p. 1. http://unstats.un.org/unsd/dnss/QAF_comments/USA%20Bur%20of%20Labor%20Performance%20Measurement_Clyde%20Tucker%20Performance%20Measurement.pdf. Accessed 25 Apr 2010.

Rodrigue J-P, Comtois C, Slack B. Conclusion: issues and challenges in transport geography. http://people.hofstra.edu/geotrans/eng/conclusion.html

Schmerbeck R. http://www.vsvi-hessen.de/download/20051102/vsvi2005_11_02schmerbeck.pdf. Accessed 26 Mar 2010.

Strassen NRW Presseinformation. www.strassen.nrw.de/service/presse/meldungen/2008/080703-01.html. Accessed 4 Apr 2010.

Strassen.NRW. www.strassen.nrw.de/service/presse/meldungen/2008/080703-01.html. Accessed 15 Apr 2010.

The glossary of federal technology. www.nextgov.com/the_basics/tb_20080605_4625.php. Accessed 31 May 2010.

The perfect performance based Contract – how should it be? www.nvfnorden.org/lisalib/getfile.aspx?itemid=607. Accessed 31 May 2010.

The project delivery system. From idea to construction. Oregon Department of Transportation. http://www.oregon.gov/ODOT/HWY/PDU/docs/pdf/PDguidebook/ProjDelSystem.pdf. Accessed 17 Mar 2010.

University of Reading. Performance-based contracting. www.icrc-reading.org/projects/projectdetails.asp?projectid=13. Accessed 11 Apr 2010.

www.embassyworld.com/maps/Maps_Of_Argentina/index.php. Accessed 18 Apr 2010.

www.general-anzeiger-bonn.de/index.php?k=loka&itemid=10001&detailid=469683. Accessed 18 Apr 2010.

www.performance-based-road-contracts.com. Accessed 23 Mar 2010.

www.reservas.net/alojamiento_hoteles/imagenes/argentina_mapaplanorutas_550x800.gif. Accessed 18 Apr 2010.

www.rundschauonline.de/servlet/OriginalContentServer?pagename=ksta/ksArtikel/Druckfassung&aid=1246895308791. Accessed 18 Apr 2010.

www.service-bw.de/zfinder-bw-web/generatepdf?type=VB&id…0. Accessed 11 May 2010.

Zietlow G. www.performance-based-road-contracts.com. Accessed 6 June 2010.

Zietsman J. www-esd.worldbank.org/pbc_resource_guide/Docs-latest edition/cases-and-pdfs/ZietsmanTexas.pdf.pdf. Accessed 10 Apr 2010.

About the Author

Ashish Gajurel is a Visiting Lecturer of Transport Economics at the Institute of Engineering, Tribhuvan University, Kathmandu, Nepal. He also teaches Logistics and Supply Chain Management at Ace Institute, affiliated to Pokhara University, Pokhara, Nepal. Besides, he is also a Transport and Traffic Consultant for several firms and organizations in the area of transportation and traffic management. He has also worked as a Supply Chain Officer at Unilever Nepal Ltd. He completed Master of Science (M.Sc.) in Transportation Systems from Technische Universität München (TUM), Germany, in 2010. His research and teaching interests include transportation and traffic engineering and management. He has authored numerous articles for national electronic and news media about traffic management, transportation infrastructure, public transportation, and so forth. As a social initiative, he also has partnered with numerous schools to conduct safety programs to inculcate safety behavior in children. He is regarded as a traffic expert in Nepal.

A. Gajurel, *Performance-Based Contracts for Road Projects: Comparative Analysis of Different Types*, DOI 10.1007/978-81-322-1302-4, © Springer India 2014

Index

A. Gajurel, *Performance-Based Contracts for Road Projects: Comparative Analysis of Different Types*, DOI 10.1007/978-81-322-1302-4,

MIX
Papier aus verantwortungsvollen Quellen
Paper from responsible sources
FSC® C105338

FSC
www.fsc.org

If you have any concerns about our products,
you can contact us on
ProductSafety@springernature.com

In case Publisher is established outside the EU,
the EU authorized representative is:
Springer Nature Customer Service Center GmbH
Europaplatz 3, 69115 Heidelberg, Germany

Printed by Libri Plureos GmbH
in Hamburg, Germany